Flexitest

An Innovative Flexibility Assessment Method

Claudio Gil Soares de Araújo, MD, PhD, FACSM
Medical Director
Clínica de Medicina do Exercício
Rio de Janeiro, RJ, Brazil

Clin MEX
Medicina do Exercício

HUMAN KINETICS

Library of Congress Cataloging-in-Publication Data

Araújo, Claudio Gil Soares de, 1956-
    Flexitest : an innovative flexibility assessment method / Claudio Gil Soares de Araújo.
        p. ; cm.
    Includes bibliographical references and index.
    ISBN 0-7360-3402-1 (soft cover)
    1. Joints--Range of motion--Measurement. I. Title.
    [DNLM: 1. Joints--physiology. 2. Range of Motion, Articular.   3. Evaluation Studies.
    4. Physical Fitness. 5. Pliability.   WE 300 A663f2004]
    QP303.A715 2004
    612.7´5--dc21

                                            2003006317

ISBN: 0-7360-3402-1

Copyright © 2004 by Claudio Gil Soares de Araújo

The Web addresses cited in this text were current as of June 2003, unless otherwise noted.

Acquisitions Editor: Loarn D. Robertson, PhD; **Developmental Editor:** Judy Park; **Assistant Editor:** Lee Alexander; **Copyeditor:** Nova Graphic Services; **Proofreader:** Sue Fetters; **Indexer:** Bobbi Swanson; **Permission Manager:** Dalene Reeder; **Graphic Designer:** Andrew Tietz; **Graphic Artist:** Kathleen Boudreau-Fuoss; **Photo Manager:** Kareema McLendon; **Cover Designer:** Robert Reuther; **Photographer (cover and interior):** Carlos Scliar; **Art Manager:** Kelly Hendren; **Illustrator:** Accurate Art, Suzana Queiroga (evaluation map artist); **Printer:** Sheridan Books; **Translator:** Carlos André Oighenstein

Printed in the United States of America   10   9   8   7   6   5   4   3   2   1

Human Kinetics
Web site: www.HumanKinetics.com

United States: Human Kinetics
P.O. Box 5076, Champaign, IL 61825-5076
800-747-4457
e-mail: humank@hkusa.com

Canada: Human Kinetics
475 Devonshire Road Unit 100, Windsor, ON N8Y 2L5
800-465-7301 (in Canada only)
e-mail: orders@hkcanada.com

Europe: Human Kinetics
107 Bradford Road, Stanningley, Leeds LS28 6AT, United Kingdom
+44 (0) 113 255 5665
e-mail: hk@hkeurope.com

Australia: Human Kinetics, 57A Price Avenue, Lower Mitcham, South Australia 5062
08 8277 1555
e-mail: liahka@senet.com.au

New Zealand: Human Kinetics
P.O. Box 105-231, Auckland Central
09-523-3462
e-mail: hkp@ihug.co.nz

Contents

Part I   Overview and History

Part II   Principles and Administration of Flexitest

Part III Research on and Applications for Flexitest

Foreword

It is an honor and a privilege for me to introduce you to this extraordinary text, which was written to fulfill two primary objectives:

- To present an improved methodology for flexibility assessment that will be adopted by major sports medicine organizations worldwide, such as the American College of Sports Medicine (ACSM)

- To have the Flexitest principles and methodology, which have been extensively employed in South America and Europe, routinely used by researchers, clinicians, and practitioners in North America

Stretching, a major component of a comprehensive physical conditioning program, is essential for increasing tendon flexibility, maintaining and improving joint range of motion and function, and enhancing muscular performance. Without a specific program to maintain flexibility, many middle-aged and older adults become hampered by stiffness in the hamstrings, shoulders, and back. Moreover, observational studies support the role of flexibility exercise using ballistic (movement), static (little or no movement), or modified proprioceptive neuromuscular facilitation techniques in the prevention and treatment of musculoskeletal injuries.

Although in recent years considerable advances have occurred relative to our understanding and appreciation of flexibility, the methodology for evaluating passive, static flexibility has remained largely stagnant since the mid-1950s. For more than two decades, a good friend and an esteemed colleague, Dr. Araújo, has been diligently working to answer a compelling question: How do we measure and evaluate flexibility? This question has relevance to health, sports, and clinical applications—and Araújo and his associates have discovered the answer. The author has meticulously developed and refined the Flexitest methodology using a research-based approach in sedentary individuals, recreational exercisers, and elite athletes, including persons with and without varied comorbidities and chronic diseases. Accordingly, this groundbreaking book is destined to become a seminal text that will serve to *change* longstanding flexibility evaluations by highlighting the advantages, validity, reliability, and ease of the Flexitest when compared with obsolete and antiquated assessment techniques (e.g., sit-and-reach test and conventional goniometry).

In addition to presenting the Flexitest and its norms, indexes, methodology, scientific foundation, and rationale, Dr. Araújo provides a comprehensive review of the flexibility measurement area, which is useful for professionals and undergraduate and graduate students in medicine, physical therapy, exercise science, physical education, athletic training, and gerontology. Especially noteworthy is the author's attention to the interpretation of flexibility data, including variability scores, a unique feature of the Flexitest that allows the tester to identify the age group of the subject being evaluated. And the good news is that the Flexitest methodology can be easily learned and performed on any given subject in less than four minutes.

This book is divided into three parts that can be independently studied. Part I provides an overview of flexibility training and testing. Part II introduces the Flexitest methodology and includes the reference charts and descriptions for evaluating the 20 joint movements used to assess flexibility, along with a method to determine the index of global flexibility, or Flexindex. Part III provides an objective, comparative appraisal of the currently available flexibility testing protocols and clinically relevant case studies. You will find figures, tables, and photographs to highlight key points. References and statistical applications support the text. Also covered are the role of flexibility in health and disease; a review of traditional flexibility testing; Flexitest methods, practices, analysis, and supporting research; a comparative table of flexibility testing methods; and the author's extensive experience in using the Flexitest with varied subjects.

This outstanding book comprises the scientific and clinical applications that will assist any trained professional in providing valid and reliable flexibility measurements for persons of any body type, fitness level, and health status. Accordingly, this text should become a standard reference for registered clinical exercise physiologists as well as physicians and their assistants, nurses, physical and occupational therapists, geriatricians, and specialists in all areas of athletic training and sports medicine.

Barry A. Franklin, PhD
Director of cardiac rehabilitation
and exercise laboratories,
William Beaumont Hospital

Professor of physiology,
Wayne State University
School of Medicine

Dr. Franklin is a former president of the American College of Sports Medicine (1999–2000) and currently serves as editor in chief for the American Journal of Medicine & Sports.

Preface

In this book we present a powerful new method called Flexitest for evaluating passive static flexibility. We have spent the past 20 years developing and perfecting this method and have used it successfully in South America and Europe. Gathered in this one concise resource for our North American colleagues are details of the principles, administration, research, and application of Flexitest.

This is not a book offering flexibility exercises! If you are looking for a guide on how to improve your flexibility or searching for new stretching routines, this is not the book for you. However, if you are interested in topics relating to flexibility and strongly believe that flexibility training is a must in all well-prescribed exercise plans, you may find information of interest in this preface. You most likely are one of the dedicated professionals working in the exercise and sport sciences field who agrees that both stretching training and good flexibility are needed by all persons, healthy or sick, young or very old, sedentary or among the athletic elite. We call your attention to a particular phrase in the preceding sentence: "good flexibility." What is this? What does it mean? How do we measure and evaluate flexibility? What is the real meaning of flexibility for health and sports? If you are really interested in exploring these questions, you are holding the right book!

It is quite surprising that in the year 2003, we still have to rely on testing procedures developed around or immediately after World War II, more than a half-century ago. Methods such as the sit-and-reach test, goniometry, the modified sit-and-reach versions, inclinometers, and so on have been introduced in the literature, but it is unlikely that they are routinely used. The situation is analogous to the somewhat dubious endeavor of repairing a very old car—you worry about the effective results, because the conceptual basis is poor. This is not because you think that flexibility evaluation isn't important, but rather because you have recognized that these methods do not provide the practical answers you need in professional daily life.

Flexibility is an important health-related physical fitness variable. However, its evaluation poses special issues. First of all, flexibility is specific to joints and movements, meaning that someone can be very flexible in the trunk and at same time display very limited ranges of motion in the shoulders or hips. Second, in contrast to other exercise parameters such as maximal aerobic power and muscle strength and power, having extremely high levels of flexibility can predispose one to injury and quite often is associated with other morbid conditions, such as mitral valve prolapse. Therefore, possessing a tool for assessing overall and specific flexibility profiles is not only desirable but, in reality, is the only way in which serious practitioners, trainers, instructors, or coaches can get a complete appraisal of flexibility. In this sense, Flexitest becomes the logical choice for flexibility evaluation.

This book is of potential use to a number of health professionals. These include physical education teachers working at the elementary and high school levels, as well as fitness instructors, sport coaches or trainers, and personal trainers working at gyms, clubs, fitness centers, and even at clients' homes. Among the many medical specialties whose practitioners may use this book and the Flexitest in their clinical practice are primary care, physical medicine, and sport medicine physicians, pediatricians, and geriatricians, as well as orthopedic surgeons. Physiotherapists and exercise physiologists also will find this book valuable, as will upper-level undergraduate and graduate physical education students and their professors, who will find the material appropriate for university-level courses. And, although it is not primarily directed at serious athletes, we believe that some of them also may be interested in learning about and being evaluated with the Flexitest.

This book is divided into three parts that can be independently studied. First, we sketch a broad overview of flexibility theory in part I, including its definitions, concepts, and terminology as well as the limiting and major factors that interfere with joint motion. Part I also covers the physiological and clinical relevance of flexibility for health and quality of life, including a discussion of hypo- and hypermobility-related disorders and an up-to-date discussion of the role of flexibility in injury prevention and incidence. Part I also includes an extensive review of the currently available classical flexibility testing methods and then presents a new, 18-point classification system.

Parts II and III, comprising six chapters, represent the core of the book. Part II introduces the Flexitest methodology and includes all the reference charts and descriptions for evaluating the 20 joint movements used to assess flexibility. It also discusses how to handle the Flexitest data statistically and determine the index of global flexibility, or Flexindex, score by adding the individual movement scores. We discuss many practical issues in flexibility testing in relation to a series of scientific data obtained with the Flexitest. The reader is schooled in the use of very important normative and percentile curves derived from male and female Flexindex data collected from over 2,600 people ranging from 5 to 88 years of age. The Flexitest results of participants of many different sports are also summarized using data from more than 400 athletes, including about a dozen Olympic medallists. The use of these normative data will permit you to really evaluate, rather than to just measure, flexibility for individuals of all age groups. With this information, you can evaluate both global flexibility and specific joints or movements, enabling you to plan a scientifically based stretching routine to achieve or maintain appropriate levels of flexibility. Readers who appreciate a good statistical approach will find in chapter 6 in-depth coverage of statistical analyses applied to Flexitest data, including the unique analysis of flexibility variability profiles.

Finally, if some uncertainty still remains about Flexitest's advantages, you will benefit by reading chapter 8 in part III, which presents the results of our in-depth and scientific comparative appraisal of the available flexibility testing protocols. There, we offer our arguments for proposing the Flexitest as the reference method in flexibility measurement and evaluation of apparently healthy subjects of all age groups and activity levels, from the very sedentary ones to the Olympic athletes. Finally, in chapter 9, case study descriptions illustrate our use of the Flexitest information and provide practical insights into the interpretation and use of the Flexitest in situations that may occur in your own professional activities.

If you are now motivated to read the book, as we expect you are, you will be surprised to find that learning to use the Flexitest is quite easy. After reading this book and learning to grade Flexitest results (chapter 5), you will be ready to evaluate flexibility by applying our statistics and techniques to Flexitest data and to use the results to further your professional progress. You will be surprised and, we hope, delighted to see how fast and easy mastering the Flexitest methodology was for you.

Additional training materials are available at the Web site of the Clínica de Medicina do Exercício (CLINIMEX) in Rio de Janeiro, Brazil, at www.clinimex.com.br. There you will find Flexitest color images and digital videos, slide shows, and simple-to-use spreadsheets that will assist you in your use of the Flexitest.

We believe that once the Flexitest method is learned and sufficiently practiced, you will find evaluating flexibility as enjoyable as we do. We also believe that your attitude and approach toward flexibility, particularly toward its measurement and assessment, will not be the same after having read this book.

Acknowledgments

To write a book in a new methodology it is not an easy task. Flexitest—an assessment tool for flexibility evaluation—was planned and delivered during my work with Gama Filho University swimming team, coached by Mr. Roberto de Carvalho Pável. I am deeply indebted to Roberto Pável, himself an enthusiast about flexibility, for bringing my attention to this interesting field and for offering his expertise and experience for the test development. Our prolonged discussions on many 1978 nights are still in my memory. The opportunity to use the Flexitest since then has significantly contributed to my professional and scientific career.

Many students and professionals contributed to the advancement and standardization of Flexitest methodology and development of an appropriate databank. Although it is impossible to cite all of them, I would like to express my gratitude, in alphabetical order, to Anselmo José Perez, Antonio Cesar Cabral de Oliveira, Antonio Claudio Lucas da Nóbrega, Astrogildo Vianna Oliveira Junior, Claudia Lucia Barros de Castro, Christianne Giesbrecht Chaves, Claudio Rebello Velloso, Karla Paula Campos, Marcos Bezerra de Almeida, Marta Inês Pereira, Paulo Cesar Haddad, and Vitor Agnew Lira. Two of my former graduate students have made additional contributions to Flexitest and to this book. Dr. Paulo de Tarso Veras Farinatti has applied, analyzed, and published Flexitest's data on children; Dr. Walace David Monteiro has helped with our reliability studies and, more recently, in selecting and grading digital photos for scoring training presented in this book and on our institutional Web site (www.clinimex.com.br).

I am pleased to thank the models for their contribution to the training photos; my flexible daughters, Aline and Claudia; my wife, Denise; my mother-in-law, Dalva Sardinha Mendes; as well as my graduate student, Aldair José de Oliveira. In addition, I must thank all subjects that agreed to be measured by Flexitest. Their contribution allowed me to derive standards for Flexitest data interpretation.

Organizing our material to publish it with Human Kinetics has been quite an experience, and I would like to thank those who worked directly with me at Human Kinetics, especially acquisitions editor Loarn Robertson and developmental editor Judy Park. Their confidence in Flexitest merits and continuous support and effort were greatly appreciated. I am also proud to have my friend and former ACSM president Dr. Barry A. Franklin write the foreword of this book.

Finally, I have to thank my family for kindly allowing me to spend hundreds of hours working on this book. I especially thank my wife, Dr. Denise Sardinha M.S. Araújo, for keeping me motivated to finish this work in the easiest and hardest times. Without her support and encouragement, I would not have completed this book. So, there is nothing more logical and scientifically valid than to dedicate this book to her.

PART I

OVERVIEW AND HISTORY

Introduction to Flexibility

One of the first things we do when we wake up, sometimes even before we open our eyes, is to stretch ourselves. Like other animals, we repeat this movement throughout the day, especially after remaining in the same position for long time. This is the simplest and most ordinary example of a flexibility exercise.

The locomotor system, with its different structures, allows the body to move. This results from comprehensive and complex actions of muscles, tendons, ligaments, and joints (Alter 1996). These actions are controlled by the central nervous system, which is instrumental in providing the broad array of the body's motor capabilities. Among this array of motor actions, some (such as dancing and running) require extreme levels of body functionality and therefore maximum performance of the locomotor system. Wide body movements are typically graceful and elegant, and they account for most of the visual beauty of dance, synchronized swimming, ice skating, and gymnastics. Performance of these movements seems to rely on a morphofunctional feature generically called *flexibility*.

Much before the creation of the word *flexibility*, Hippocrates described individuals from a specific ethnic group of his time who presented with an exaggerated joint laxity that prevented them from throwing javelins without injuring themselves (Grahame 1971). Medical syndromes characterized by excessive joint mobility were first described in 1892 by Tschernogobow in Russia (Ehlers-Danlos syndrome) and by Marfan in 1896 (Marfan syndrome) (Grahame 1971).

The word *flexibility* is not new in literature or use. It is likely that it derives from a mixture of the words *flexion* and *capability*. One of its first uses was for describing toe touching with the arms fully extended, either standing or sit-ting, with anterior trunk flexion and the legs kept straight (Cureton 1941). Kinesiologically, flexion is not the only possible movement—it is also possible to perform extension, adduction, and abduction in the body joints. But nonetheless, the original association of the word *flexion* remains in the term *flexibility*.

Flexibility has recently been included as a fundamental variable in exercises for healthy adults (ACSM 1998a) and the elderly (ACSM 1998b). Although flexibility exercises are included in a comprehensive exercise prescription, it is interesting to note that there is much less (and much less current) scientific documentation published on this matter than on other major health-related physical fitness variables, such as maximal aerobic power, muscular strength and endurance, and body composition. Thus, it is not surprising that discussion of the assessment and prescription of flexibility exercises is often general. The scarcity of data could be due to major limitations in the assessment techniques employed, particularly in terms of their validity and lack of an established index or score for individual global flexibility. These limitations have also hampered comparisons of flexibility in individuals of different age groups or genders and of those who practice different sports, as well as the efficacy of specific training programs on flexibility.

This chapter reviews the general and most relevant aspects of flexibility, presenting definitions, the role of specificity and physical training, and the influence and relevance of intervening factors in flexibility.

Definitions

As mentioned, professionals in the field of physical activity have been using the term *flexibility* for quite some time. Most individuals, even laypersons, have an idea of the meaning of the word and even of some of its implications both in physical training and fitness and in health. Because of the broad concepts and applications of flexibility, presentation of a new flexibility assessment technique should address the different definitions of the word and, ideally, introduce a new, comprehensive definition.

The simplest definition considers flexibility as the "range of motion of a joint" (Stoedefalke 1974; Mathews 1978). Phillips and Hornak (1979) added to the definition "or sequence of joints." Concern over the strictly physiological measurement was expressed by Reilly (1981), who introduced "lack of rigidity" to the classical definition of flexibility, and by Bosco and Gustafson (1983), who defined flexibility as "the range of motion of body parts over their joints, without excessive stress on these joints or their tendons and ligaments." Definitions more recently proposed incorporate the "maximum range of motion" (Kell, Bell, and Quinney 2001), wherein flexibility is presented as "the ability of a joint to move over the entire range of its movement arc" (Fahey, Insel, and Roth 1999), which, in reality, simply details the meaning of "maximum range of motion." This definition is very similar to the one proposed in a recent ACSM position statement (1998a).

> Different authors have defined flexibility differently, but all of the definitions included the expression "range of joint (or joints) motion."

A fundamental issue in defining flexibility is clarifying how the maximum range of motion is quantified. The maximum limit of a particular joint movement can be reached *actively*, by a person contracting his own muscles, or *passively*, with assistance in moving the joint or limb being provided by another person. Because the passive range of motion is often larger than the active, as well as being less influenced by other variables (such as muscle strength and coordination), it is normally preferred for evaluating flexibility. Consider, for example, knee flexion from a standing position; the range of motion obtained by hamstring muscle contraction is limited by the location of muscular insertions. However, with the help of another person, it is quite possible to increase the range of motion, often to the point of superposing the posterior portions of the leg and thigh and achieving a substantially wider arc of movement. A more extreme example is that of a person with paraplegia who has no active ankle flexibility but whose ankle range of motion—dorsiflexion and plantar flexion—is normal or seminormal on passive mobilization.

The arc of movement of a given joint is limited by structural factors and by the discomfort experienced by the person being assessed. Both of these factors vary considerably among individuals and even among different joint movements for the same person. Depending on the joint range of motion that is being assessed, different limiting factors can restrict the range. In addition, there are some movements for which achieving the maximum amplitude could generate a high degree of discomfort (i.e., lateral shoulder rotation or wrist flexion), whereas for others (i.e., wrist extension or elbow flexion) this is not the case. Therefore, to avoid causing injury during flexibility measurement, one must consider the physiological rather than the potentially pathological range of motion.

Range of motion is a specific rather than a generic feature, and it is possible that one person may have a wider arc of movement for some joints and a more restricted one for others. This characteristic has been incorporated into the definition of flexibility in a broader sense as "specificity of flexibility" (Dickinson 1968; Harris 1969a). As for the static component of joint mobility, it can be noted that the available definitions of flexibility do not address some of the important aspects we have just discussed.

A definition of flexibility must therefore account for several important points. First, it should explicitly incorporate the idea of *maximum range of motion*, as previous definitions have indicated. It should also encompass *passive measurement* as a means of eliminating or minimizing the influence of other variables such as muscular strength, motor coordination, and motivation of an individual in measuring range of motion. Furthermore, it should explicitly state the need to prevent injury when measuring the maximum range of motion by using the term *physiological* to stipulate that the maximum range of motion should be reached without injuring tissues or joints. Finally, our definition should meet the requirement of *specificity* by stating that flexibility is measured for a given joint movement. Therefore, our definition of flexibility is as follows:

> *The maximum physiological*
> *passive range of motion*
> *of a given joint movement*

It is important to mention that this definition emphasizes movement of individual joints rather than overall flexibility. However, the morphofunctional feature called flexibility tends to be characteristic for the body as a whole and is not specific to only a given joint movement. In fact, both aspects—the range of motion in individual movements and overall or global flexibility (the latter of which is difficult to define)—are extremely relevant.

A final word about the definitions of other flexibility-related terms: Though we realize that *joint mobility* may be used as a synonym for *flexibility* (if one considers only the static component) (Leighton 1955), the word *stretching* is better employed to indicate a type of physical exercise in which one reaches a maximum or quasi-maximum range of motion, actively or passively, for one or more joint movements. Therefore, we understand that it is correct to use the expression *stretching exercises* as a synonym for *flexibility exercises.*

Specificity

One of the fundamental questions in flexibility research is whether joint mobility is a general feature of an individual. Cureton (1941), in a classical paper, devised four tests for measuring and assessing flexibility. That the correlation of these four scores was quite low suggested the tests' specificity (i.e., each test measured slightly different features and therefore was not indicative of overall flexibility). Despite the theoretical limitations of the tests, which allowed variables other than flexibility to influence the result, this classic study called atten-

tion to the fundamental issue of the specificity of flexibility.

The issue of specificity was neglected for many years, perhaps because of the widespread use of single-movement linear tests in the 1950s and 1960s, but it was readdressed later (Dickinson 1968; Harris 1969a). Dickinson's pioneering work (1968) showed that there was no significant correlation between flexion and extension mobility for the wrist and ankle, and put forth the theory of specificity of flexibility. This work was complemented later by that of Harris, who suggested that flexibility is not a generic feature of the human body. Thus, Harris concluded, no single joint movement measurement or single test that combined results of a few movements and presented them as a single result could satisfactorily represent a given individual's flexibility characteristics (Harris 1969a; Harris 1969b). Harris (1969b) investigated 147 right-handed female University of Wisconsin undergraduate students and, by applying factorial analysis (a quite advanced statistical assessment for the time), showed that static and dynamic flexibility were two different features and also that specificity was not restricted to joints, but also applied to movements of a joint, which could have substantially different ranges of motion. For example, an individual may present good flexion and poor extension at the same hip. Thus, it is difficult to consider a person or even a joint flexible unless broad and comprehensive joint mobility measurements are made.

Relevance

Once flexibility is presented, defined, and delimited, it is necessary to discuss the meaning and relevance of joint mobility and therefore of its measurement. In health, particularly when considering the elderly or people with physical disabilities, autonomy, independence, and safety (for instance, minimizing the risk of falls) depend on the person having the appropriate levels of body flexibility in general, and in particular on the range of some joint movements (Gersten et al. 1970; Schenkman, Morey, and Kuchibhatla 2000; Hauer et al. 2001). Flexibility also plays a fundamental role in some sports and performing arts. Even for musicians, giving an outstanding performance depends on the mobility of a number of joints. On the other hand, there is no relationship between flexibility, as assessed by trunk flexion mobility, and all-cause mortality (Katzmarzyk and Craig 2002).

Regular exercises for maintaining and enhancing joint mobility have been recommended by practically all institutional position statements on physical activity (ACSM 2000; Pollock et al. 2000). Interest in flexibility was enhanced when it was included as a physical fitness variable in the 1950s (Corbin and Noble 1980). More recently, that interest was renewed with the acknowledgment that flexibility is a component of health-related physical fitness (Bouchard et al. 1990). Therefore, quantifying flexibility—from the lack of it to its excess—is indeed relevant. These aspects will be further discussed in chapter 2. In addition, the study and quantification of flexibility are also relevant to areas of knowledge ranging from biomedical basic disciplines to applied sciences.

▶ Flexibility Is Relevant to Many Areas of Knowledge

- ▼ Biophysics
- ▼ Physiology
- ▼ Kinanthropometry
- ▼ Exercise and sports medicine
- ▼ Physical education
- ▼ Physiotherapy
- ▼ Orthopedics and traumatology
- ▼ Physical medicine
- ▼ Rheumatology
- ▼ Ergonomics
- ▼ Engineering

Limiting Factors

There are several major physical restraining factors for joint motion, which we will discuss in this section (table 1.1). For practical reasons, we will not consider paresis as a limiting factor. Those interested in obtaining more details

Table 1.1	Flexibility Limiting Factors
Articular	**Capsular**
Connective tissue	Synovial liquid Tendons
Muscles	Ligaments Connective components Viscoelastic components
Fat	Hypertrophy Subcutis
Bones	Viscera
Skin	

should read other texts that discuss joint mobility from a deeper anatomical and histological perspective (Holland 1968; Kapandji and Kandel 1997; Alter 1996).

Joint mobility results from the action of a force over body segments connected by a joint. If this force derives from muscular contraction, an *active* movement is being performed; if this force is external to the body, whether caused directly or indirectly by gravity or by another individual or object, the movement is termed *passive*. Joint mobility, from an anatomic, kinesiological, and physiological point of view, is a continuous and finite trait that ranges from immobility to an extraordinarily wide range of motion.

The principal body joints are morphologically and functionally related to the different tissues and structures, including articular capsules, tendons, ligaments, muscles, fat, bones, and skin. There are, however, a number of operational difficulties in determining the role of each of these in limiting in vivo joint mobility. Furthermore, for each movement and joint, the limiting factors may be different. In addition, during an individual's lifetime, it is quite likely that the relative role of each factor will change.

Johns and Wright (1962) objectively quantified the relative importance of several factors in the physical restraint of a given joint movement. The properties of a cat's wrist (theoretically similar to the human metacarpophalangeal joint) were carefully investigated to quantify each one's role in resistance to movement by selectively dissecting each restraining component and measuring the torque. The results revealed that inertia and viscosity represented less than 10% of resistance, whereas plasticity and particularly elasticity were the main factors to be overcome in the performance of a movement. From an anatomical point of view, the articular capsule was responsible for 47% of the resistance; the muscle, for 41%; the tendon, for 10%; and the skin, for only 2%. However, it should be noted that these measurements were made in movements that only went up to half of the joint's maximum passive range of motion. As suggested by these authors, in the extreme point of joint amplitude the tendons have a much more important role, particularly in a movement such as wrist flexion.

While this detailed study provided a significant contribution to scientific knowledge, one should expect variability in these proportions when assessing a person for various joints and movements and when considering different age groups, genders, and physical characteristics. For instance, it is quite common to observe a mechanical limitation in trunk flexion caused by an excess of abdominal fat or advanced pregnancy.

Articular structures also play a role in restricting the arc of movement caused by capsular and extracapsular factors that become considerably more important in some morbid states, such as those seen in different forms of joint diseases. In some pathological states, joint inflammatory responses featuring flogistic signs (pain, edema, redness, and so on) may account for considerable limitation of mobility (Williams 1957); the same is true for major fibroid scars (Wilson and Stasch 1945). This type of mobility restriction is most often seen in patients with rheumatoid arthritis or related diseases.

A considerable amount of the resistance to a movement is due to the major proteins in the muscle (actin, myosin, dystrophin, and others), the muscle-connective tissue interface (laminin, for example), and the connective tissue itself. Of particular interest is that the extreme range of the arc of movement depends on the relative proportions of collagen and elastin proteins in connective tissue; because elastin is easily distensible, collagen is the primary cause of restraint.

When we consider the skeletal muscle as a joint-mobility-restraining factor, it is important to specify the conditions of the study. Although skeletal muscle has little role in the passive motion of our fingers' small joints, it can be a limiting factor for trunk flexion when the hamstrings present an augmented tonus and a shortened resting position. In general, restraint due to muscles tends to be more evident in major joints. This is more easily noted in people with extreme muscular development, such as bodybuilders, in whom complete elbow extension is rarely achieved.

Sapega and Nicholas (1981) stated that the main causes of physical resistance to muscular stretching were the connective components that define the muscular skeleton, not the contractile ones. According to Barnett and Cobbold (1969), elastic tension accounts for at least half of the muscular resistance to movement. Obviously, abnormal spastic and muscular contraction states may considerably increase resistance to muscular stretching and could even completely prevent joint movement.

Other structural factors may also be important in limiting joint motion. One of these factors is the amount of fat tissue (Reilly 1981). Distribution of body fat is initially centrifugal, but becomes centripetal with body growth and development. This is why one can note in neonates a limited arc of movement in wrist extension, caused primarily by excessive fat tissue deposition at that site, whereas an obese adult may find it difficult to perform knee flexion or hip adduction because of the mechanical obstruction caused by the excessive amount of fat collected at those locations.

Bone position in a joint may also represent a basic restraining mechanism. For instance, it is extremely difficult, if not impossible, to perform elbow adduction or abduction. The socket between the humerus and forearm bones practically prevents these movements from being made. Bone sockets—more precisely, the olecranons at the end of the olecranian fossulae—limit elbow extension. It is even possible that the range of motion for this movement cannot be augmented after epiphyses closure (Reilly 1981). In certain circumstances, especially for a large arc of movement, direct bone superposition will also limit the range of motion, as it could in elbow flexion.

Finally, one must consider skin, which typically has a minor role in restraining mobility, but in some pathologic conditions may have augmented or decreased elasticity. For instance, in the cases of dehydration and Ehlers-Danlos syndrome, the skin is even less restraining, whereas in scleroderma and other clinical conditions of skin thickening, its restraining qualities may be enhanced.

Intervening Variables

Academic pursuits are prone to the formation of opposite opinions and contradictions, and this rule also applies to the study of flexibility. Most discrepancies are due to researchers' using different methods to evaluate flexibility or analyzing different populations. Joint mobility may be influenced by a number of aspects, including age, gender, morphologic features, and the regularity of physical exercise. This section reviews relationships between flexibility and other intervening variables (with our aim being to have a more practical than academic approach), based on the available literature and our own experience and viewpoint.

Age

In 1921, Gilliland suggested that perhaps different joint mobility values should be considered for children and the elderly. This idea has been subjected to a number of investigations.

Coon et al. (1975) established standards for six-week-old and three- and six-month-old babies by passively measuring the angles of the knees of more than 40 children. They noted that knee mobility was somewhat higher in three- and six-month-old babies than in those of six weeks of age.

Haas, Epps, and Adams (1973), Hoffer (1980), and Waugh et al. (1983) also studied mobility in newborn babies and found that it progressively increased in the upper limbs but stayed constant in the lower limbs over

the first three days of life. Interestingly, ankle plantar flexion gradually increased, whereas dorsiflexion decreased. Hoffer (1980) also noted that a limitation on knee extension, which may be 35° degrees at birth, tends to disappear only in the early stages of gait in the second year of life. These data suggest the existence of a specific joint mobility pattern in newborns that somewhat reflects a mobilization pattern and the typical intrauterine position. It is also interesting to note that there is no evidence of ligamentous hyperlaxity or hypermobility in the first week of life (Wynne-Davies 1971).

There is limited scientific information about flexibility from six months to five years of age. This is likely because of difficulties in investigating children in this age group, and also because of a lack of interest by the medical and sports disciplines. Through the use of some ligamentous laxity measurements, Wynne-Davies (1971) noted that most children of age two or three can be considered hypermobile; from then on, there is a steep loss in the extreme mobility level. Similar data were also obtained in a classical study by Beighton, Solomon, and Soskolne (1973).

Flexibility evaluation techniques show that joint mobility is maintained or gradually decreases over the years of childhood and adolescence (Gurewitsch and O'Neill 1941; Kendall and Kendall 1948; Leighton 1956; Silverman et al. 1975; Lehnhard et al. 1992; Farinatti, Nóbrega, and Araújo 1998), even though distinctive patterns may occur in some joints (Leighton 1956) or in specific movements (Goldberg et al. 1980). When analyzing flexibility information obtained by administering the Flexitest to 901 boys and girls between 5 and 15 years of age, Farinatti, Nóbrega, and Araújo (1998) noted a tendency for overall flexibility to decrease and, more specifically, for diminishing trunk mobility. In opposition to the general rule, Boone and Azen (1979) noted a peculiar feature: a higher degree of wrist and elbow extension in children between 5 and 10 years of age compared to children in their first 5 years of life.

In contrast to the rather small number of studies on flexibility during the years of growth and development, there are many more on flexibility in adults. As expected, it has been observed that there is a gradual decrease of joint mobility over the years, regardless of the assessment technique employed (Kottke and Mundale 1959; Macrae and Wright 1969; Moll and Wright 1971; Allander et al. 1974; Sugahara et al. 1981; Einkauf et al. 1987; Shephard, Berridge, and Montelpare 1990; Brown and Miller 1998). Despite this overall decrease in flexibility over the years for both males and females, the rate of loss seems to differ among different joints and depend on how flexibility is measured. According to Boone and Azen (1979), differences between children and adults are more evident in the shoulder lateral rotation range. Smahel (1975) found that differences between active and passive flexibility tend to decrease over the years.

Although there is a considerable amount of data pointing to major flexibility loss with age, two relevant issues still have no accurate or definite answers. First, what is the magnitude or amount of loss of flexibility that occurs with aging? Second, what are the causes of the age-induced reduction in flexibility? Most evaluation methods or techniques do not supply an objective way to evaluate loss of flexibility with aging, and currently there are no appropriate long-term longitudinal data available. Data gathered to assess ligamentous hyperlaxity make it clear that the prevalence of this condition is quite stable in the childhood and then falls quickly and steeply in adulthood, going from 50% to under 5%, and from then on stabilizes or lowers slightly until old age (Beighton, Solomon, and Soskolne 1973). While flexibility is reduced in the later years of adult life, it is clear that ligamentous laxity assessment does not allow an accurate quantitative or qualitative analysis of overall joint mobility behavior (this issue will be further addressed in chapter 8). Recent data suggest that flexibility may be largely stable from adolescence to young adulthood (Lefevre et al. 2000; Fortier et al. 2001). From a study using Flexitest methodology (see pages 49-76), we were able to observe two key points:

1. It is clear that there is a substantial loss in flexibility between ages 5 and 80; elderly subjects present about half of the overall flexibility scores observed during childhood.

2. A small group of family-related women showed some flexibility stability and large intersubject variability when reevaluated 15 years later.

Considering these preliminary data, it seems worthwhile to acquire long-term longitudinal flexibility data in order to clarify the issue of flexibility loss with aging.

During development, muscular mass is gradually gained to support complex daily motor activities (for instance, walking, playing, and jumping). Later in life, the lessening elasticity of connective tissue structures, partially due to higher collagen crystallinity and increased fiber diameter, accounts for the expected gradual reduction in body flexibility. Further understanding of the physiological and biochemical aging processes in these tissues may contribute to the establishment of strategies for maintaining the optimal flexibility levels of childhood and adolescence.

▶ Flexibility and Age

▼ Trend for decreasing flexibility with the aging process

▼ Maximum values tend to be achieved during early childhood

▼ Interindividual variability also increases with aging

In summary of the literature, and in light of our own research using Flexitest methodology, it is clear that the excessive flexor tonus of newborns limits their mobility in the first hours of life. As weeks and months pass, the tonus balances, and maximum flexibility is achieved around two or three years of age. From then on, flexibility tends to continuously decrease until old age. It seems that the velocity of joint mobility loss is not constant during the aging process. The loss seems to be faster from childhood to the end of adolescence, much slower in the following two decades, and again faster in the following years. It also seems clear that the reasons for this loss may differ for each movement and joint, although this issue has not been subjected to an appropriate prospective study. In our data, we have noted a tendency for better mobility preservation of distal as compared to proximal movements with aging. We have also seen greater variability in overall flexibility in middle-aged adults when compared to children, adolescents, and young adults. In our opinion, this happens because younger people tend to be more uniformly active (for instance, by attending mandatory physical education classes), whereas physical activity levels in older adults range more broadly, from a total lack of exercise to very frequent and intense exercise and sports practice. Therefore, it is clear that there is a need to develop flexibility evaluation norms for each age group. It is also essential that large longitudinal studies be performed so that current knowledge, which is based almost entirely on transversal studies, may be confirmed or disputed.

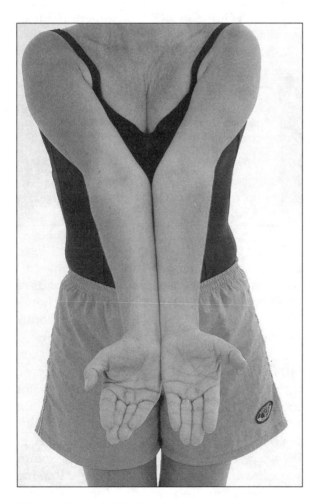

Figure 1.1   Cubital angle.

Gender

At the end of the 19th century, Potter (1895) showed that males and females differ in terms of *cubital angle* (figure 1.1). Since then, joint mobility differences in males and females have been examined in a number of studies, with different conclusions. Here we present the results of key studies, analyzing movements in a proximal–distal or cephalocaudal direction.

The first major area to be considered is the temporomandibular joint. It has a major role in a number of actions (such as talking, eating, drinking, and kissing), and it is the only joint in which males systematically have wider opening or greater flexibility (Wright and Hopkins 1982).

Proceeding to the upper limbs, active mobility for lateral and medial shoulder rotation has been found to be slightly better in females (Murray et al. 1985). Many authors have found in the past that females have a larger cubital angle as compared to males (Keats et al. 1966; Baughman et al. 1974; Beals 1976). By employing statistical techniques unavailable in the 19th century, they found that flexibility outcomes of males and females were highly superposed and frequently without significant differences, but that females tended to have slightly better results. As for wrist mobility, Smahel (1975) and Bird and Stowe (1982) found higher mobility in females, although the difference was less evident in elderly subjects. In the fingers of the hand, there seemed to be no significant gender difference (Cantrell and Fisher 1982).

Depending on the evaluation technique employed, investigations on gender differences in trunk flexibility report varied findings. Moll and Wright (1971), employing a linear technique, found higher flexibility for males—presumably because they tend to be taller. The influence of technique was reported by Wolf and colleagues (1979), who reported that males had higher *mobility* when the linear method, which measures in centimeters, was used, whereas females had higher *amplitude* when goniometry was employed. When trunk flexion alone was analyzed with the Kraus-Weber technique, a number of authors found better results for girls than for boys (Phillips et al. 1955; Kelliher 1960). These results were confirmed by Grana and Moretz (1978), who employed the Nicholas flexibility assessment technique and found similar results. These results prompted Goldberg et al. (1980) to propose for this technique different evaluation guidelines based on gender.

In the lower limbs, women tend to be more flexible, although the ankle demonstrates a decreased gender difference (Nowak 1972). Although women tend to present higher plantar flexion, it is fairly common for them, particularly after 40 years of age, to show a steep reduction in the ankle dorsiflexion range of motion when compared to men. This is likely associated with women's frequent use of high-heeled shoes in Western countries (Alexander et al. 1982).

In a review of overall flexibility, two cross-sectional population studies found similar results and a uniform tendency for females to have higher flexibility in practically all age groups (Beighton, Solomon, and Soskolne 1973; Allander et al. 1974). Our data, obtained using the Flexitest, show that gender differences are minimal in five- or six-year-old children (see chapter 6, pages 111-136). However, from this age forward a gender gap is apparent—especially after puberty—with women always presenting better results.

> From as early as six years of age, most females are more flexible than age-matched males, although there is a considerable overlap between their data distributions.

In short, there is general consensus that women, from at least the beginning of formal schooling, are more flexible than men in all joints except the temporomandibular joint. It is also clear that the evaluation technique may influence the magnitude of the differences and even, in rare cases, result in atypical results. Also, because values for both genders are considerably superposed, it is certainly possible for a man to be more flexible than a woman of similar age. The biological reasons for these gender-based differences in joint mobility are still unclear, and methodological complex-

ity will probably keep it that way. We may hypothesize that hormonal (i.e., high serum levels of relaxin during pregnancy), cultural (i.e., the expected graceful female style), and morphological (i.e., smaller muscular tonus and more lax ligaments) factors play a role in these differences.

Laterality

Joint mobility measurements are routinely taken on only one side of the body in the assumption that, in subjects who are not physically disabled, the two sides of the body have identical mobility levels. A large majority of studies have found that mobility of the right and left sides of the body are highly similar. Glanville and Kreezer (1937) were perhaps the first to report that in adults, there was no difference in mobility measures for the two sides of the body. This finding was further validated for different movements (Allander et al. 1974; Boone and Azen 1979); for joints of the wrist (Nemethi 1953) and ankle (Alexander et al. 1982); and for lateral cervical flexion (Ferlic 1962) and trunk lateral flexion (Rezende, Faria, and Almeida 1981). Yet occasionally, one may identify a few cases of bilateral differences.

While these findings of lateral similarity are the rule for ordinary people, athletes may present significant asymmetric features, particularly those whose modalities require repeated unilateral efforts. For instance, Chinn, Priest, and Kent (1974) found that upper limb joint mobility in tennis players is considerably asymmetrical, and Kirby et al. (1984) identified a higher incidence of bilateral differences in athletes compared to nonathletes. In soccer, a sport that does not favor a particular side of the body, Oberg et al. (1984) found no significant flexibility differences between the two sides of players' bodies.

The assessment of the influence of *laterality over mobility* also takes into account other factors, such as the evaluation technique selected. This may be exemplified in data from Smahel (1975), who identified the same range of motion on the sagittal plane of both wrists, and yet extension was higher in the right wrist and flexion was higher in the left.

> The arcs of movement in some joints may differ according to lateral dominance or preferential use.

After reviewing the literature and considering our own experience in evaluating flexibility, it seems likely that joint mobility is basically identical for the two sides of the body in healthy subjects who are not athletes or do not have a predominantly "unilateral" occupation. This conclusion is substantiated in practice by the medical strategy of assessing the mobility of an injured joint by comparing it with the mobility of its contralateral joint (Moore 1949b; Stoedefalke 1974; Rusk 1977; Roaas and Andersson 1982).

Morphologic Factors

Flexibility, a variable of morphologic and functional features, may be classified in the academic field of kinanthropometry. Thus, one should review flexibility's relationship to other aspects of body structure and function, such as anthropometric measurements, somatotype, body composition, muscular strength, and power. This section will review these interrelationships.

The relation between flexibility and body height was addressed in past decades in a number of studies, all of which pointed to a lack of significant correlation between the two variables (Mathews, Shaw, and Bohnen 1957; Anderson and Sweetman 1975; Dockerty and Bell 1985). Because these authors studied samples with relatively uniform age groups, the association between gradual losses in height and flexibility due to aging did not interfere with the results.

In most studies in the 1950s and 1960s that involved using linear measures for assessing joint mobility, a major concern of researchers was to identify associations between flexibility and anthropometry. However, no significant relationships could be found between flexibility and selected anthropometric measurements (Broer and Galles 1958; Mathews, Shaw, and Woods 1959; Burley, Dobell, and Farrell 1961).

Among morphologic variables, only body composition and flexibility are regularly incorporated into health-related physical fitness appraisals (Bouchard et al. 1990). The simplest strategy for an overall morphologic assessment is to calculate a person's body mass index (BMI), which is the ratio between body weight (in kilograms) and the square of height (in meters). While the large body of epidemiological documentation on BMI has allowed for the establishment of normal clinical ranges that are associated with lower morbidity and mortality rates, there are a number of theoretical and practical limitations to this index, such as the absence of specific values for children and adolescents. Also, the fact that one cannot differentiate between the causes for high BMI values—i.e., whether the excess mass is fat or muscle tissue in overweight subjects—and that interpretation is compromised when height is much over 170 centimeters (Ricardo and Araújo 2002) point to limitations of using BMI alone to test the association between flexibility and body composition. These limitations may explain why a study of Dutch children found no relationship between flexibility and BMI (Rikken-Bultmann, Wellink, and van Dongen 1997).

In the sports and exercise field, body composition is often analyzed with a two-component model, lean body mass and fat mass. Despite some limitations in different methodologies and predictive equations, it seems clear that there is a range of body fat percentage that reflects appropriate levels of fitness and health. One of the strategies used to assess body composition, and at the same time to gather information about a subject's relative linearity, is *somatotypology*. This concept was first introduced in the 1940s and became more broadly applicable when the anthropometric technique developed by Heath and Carter (Heath and Carter 1967; Carter 1970) was applied in populations of elite athletes. In short, somatotype is expressed by three components to which independent numeric values are ascribed so that relative predominance can be identified (Ross et al. 1979):

1. Endomorphy, or body fat
2. Mesomorphy, or level of musculoskeletal development related to height
3. Ectomorphy, which expresses relative linearity and is calculated according to a reciprocal of the ponderal index

Several studies have attempted to associate flexibility with somatotype, but no significant relationship was found (Tyrance 1958; Laubach and McConville 1966; Beighton, Solomon, and Soskolne 1973). Our experience with the Flexitest has confirmed this fact, although some generic trends can be noted. In extremely obese subjects, the excess of fat limits range of motion in some joint movements; the same trend is seen in individuals with excessively developed muscles. On the other hand, predominantly ectomorphic subjects, especially women, tend to exhibit higher general flexibility and ligamentous laxity values. This may be related more to a specific genetic profile than to an objective and causal association between relative linearity and flexibility.

When one considers the relationship between muscular strength or power and flexibility, it brings to mind competitive bodybuilders who have exceptional levels of muscular development and, quite often, very modest levels of body flexibility (Chang, Buschbacher, and Edilich 1988). Recent cross-sectional data have shown that bodybuilders showed 10% less shoulder rotation range of motion than age-matched non-bodybuilders (Barlow et al. 2002). In the literature, there are some reports of contradictory results from longitudinal studies in which flexibility was measured pre- and post-resistance training. It is certainly possible to enhance muscular strength without significantly reducing joint mobility in a subject with a normal flexibility level if a well-structured exercise program is followed (Massey and Chaudet 1956; Kusinitz and Keeney 1958; De Vries 1974). In this sense, it seems worthwhile to comment on the recent findings of simultaneous gains in strength and flexibility in previously inactive older adults (Fatouros et al. 2002). However, it is also true that inadequate exercise may reduce joint mobility levels (Watson 1981), particularly when femoral quadriceps strength is increased (Moller, Oberg, and Gillquist 1985). In hypermobile subjects, an increase of muscular mass tends to reduce the range of joint motion, thus minimizing the nefarious late consequences of hypermobility. Similarly, acute muscular hypertonicity, whether due to exhaustion from just-completed physical exercise, delayed muscle soreness from exercise sessions on previous days, or even intense anxiety, may lead to a reduction in the amplitude of a joint movement.

> There are some modest relationships between flexibility and other morphological factors, including a tendency for greater flexibility in taller and predominantly ectomorphic individuals and for lower levels of joint mobility in endomorphic and mesomorphic individuals.

It may be concluded that the main morphologic variables, as currently measured and assessed, modestly relate to flexibility. However, this topic has not yet been exhausted, and it is possible that future studies may partially or totally change the conclusions presented here.

Physical Training

Flexibility changes in response to specific training. There are a number of books that describe methods, techniques, and appropriate strategies for flexibility training (Alter 1998; McAtee 1999). Aerobic exercises that are not directly related to flexibility, whether performed in water or on land, tend not to induce gains in the ranges of motion of joints (Taunton et al. 1996). To enhance flexibility, specific exercise programs that typically use stretching routines for the principal joint movements are needed. Many exercises combine two or more joint movements to optimize training time, which seldom takes more than a few minutes two or three times a week (Alter 1998; ACSM 2000).

Ballistic exercises were initially used to increase flexibility, followed by techniques that alternated cycles of contraction and relaxation; their techniques were based on the theory of proprioceptive neuronal facilitation (Burke, Culligan, and Holt 2000). More recently, predominantly static forms have been used. In these, a subject reaches an extreme position (typically when she starts to feel some discomfort) and remains in that position for a length of time ranging from 15 to 90 seconds, with 30 seconds being considered optimal (Bandy and Irion 1994; Bandy, Irion, and Briggler 1998; Feland et al. 2001). Unfortunately, most studies have involved very short periods of training, rarely of more than one year in length (Morey et al. 1996).

While there is a consensus that flexibility exercises should be prescribed for healthy and unhealthy subjects and for athletes and dancers (ACSM 2000), the biological mechanisms responsible for the favorable effects of specific training are relatively unknown. In the 1980s, the emphasis was on the neurophysiological mechanisms associated with the different forms of exercises (Moore and Hutton 1980; Sady, Wortman, and Blanke 1982). More recent papers show that research is focusing on muscle and tendon viscoelastic properties (Sapega 1981; Etnyre and Abraham 1988; Magnusson 1998; McHugh et al. 1998; Kubo et al. 2001; Kubo, Kanehisa, and Fukunaga 2002) and on cellular mechanisms (De Deyne 2001) rather than on neural theories.

In carefully developed research, Magnusson has shown that the principal mechanism associated with acute and chronic increases in range of joint motion with specific training is a *higher stretch tolerance*, with no substantial changes in muscle viscoelastic properties (Magnusson et al. 1996; Magnusson 1998). Corresponding research has shown that a single five-minute session of specific stretching induces an increase in the range of hamstring extensibility without affecting the properties of muscle stiffness curves; this change is caused solely by the increased tolerance to stretching (Halbertsma, van Bolhuis, and Göeken 1996). Interestingly, Magnusson (1998) reported that the acute effects of enhanced joint mobility induced by a stretching session disappeared in less than one hour. Notwithstanding these recent and intriguing data, Magnusson (1998) does not rule out the possibility that specific training over longer periods of time, as dancers and athletes do, may lead to distinct types of chronic adaptations. More recently, relevant information about the *in vivo* behavior of tendons during stretching exercises was obtained, and it may be useful for gaining a better understanding of the biological mechanisms involved in the training of flexibility (Kubo et al. 2001; Kubo, Kanehisa, and Fukunaga 2002).

> Recent data show that one of the main reasons training induces an increased range of motion is that it increases stretch tolerance.

Regardless of the biological mechanisms responsible for joint range of motion adaptations to stretching exercises, most studies show that there is significant flexibility improvement with training programs of just a few weeks' duration in which there are just a few minutes of effective stretching exercises.

External and Other Factors

Because warming up before exercise triggers a number of physiological responses, it is appropriate to review the relationship between flexibility and temperature. The simple raising of body temperature artificially by external means, such as by going to a sauna, does not seem to cause any major increase in body flexibility. However, with physical exercise, there is an increase in joint and muscle temperature. For instance, Hamilton (1967) showed that there was a substantial increase in the proximal interphalangic joint range both with passive exercise and by combining exercise and application of local heat. The simple repetition of the exercise a number of times leads to an increase in mobility for that specific movement (Fieldman 1966; Atha and Wheatley 1976; Frost, Stuckey, and Dorman 1982; O'Driscoll and Tomenson 1982), even though this is not necessarily related to an increase in the temperature at the site. Atha and Wheatley (1976) had a group of subjects perform 20 trunk flexions from a seated position and measured the reach of their arms. They observed a 4-cm improvement between the 1st and 10th repetitions, with a proportionally higher gain in the initial repetitions and stability near later repetitions, suggesting that the maximum effect had been achieved.

Overall physical activity may lead to an acute increase in joint mobility. Hubley, Kozey, and Stanish (1984) reported that static or dynamic exercises led to greater hip mobility. The favorable effects of a stretching session lasted for at least 90 minutes (Moller et al. 1985a). Interestingly, there is evidence that warm-up may further enhance the existing joint mobility differences between athletes and nonathletes (Kirby et al. 1984) and between the "trained" and "untrained" arms of tennis players (Chinn, Priest, and Kent 1974).

In summary, in individuals with regularly moving joints, *active warm-up*, i.e., that done by voluntary muscle contraction, may enhance flexibility, whereas *physical agents* that increase body temperature, such as local hot packs or sauna exposure, tend not to be very effective. Therefore, before carrying out a flexibility evaluation, the intensity, duration, and characteristics of previous physical activity should be carefully quantified and controlled.

> Flexibility may be enhanced by exercise (active) or by physical or external factors (passive) able to induce increases in joint or body temperatures.

Theoretically, it may be presumed that other variables can influence or depend on flexibility. It is thought that the maximum range of joint movement may be genetically influenced, even though it is theoretically and methodologically hard to quantify this issue. Recent data (Katzmarzyk et al. 2001) suggest that the genetic component in trunk flexion range is 64%, higher than the genetic component observed in tests of muscle strength and endurance. Furthermore, these authors (Katzmarzyk et al. 2001) have observed an important role

played by heredity on flexibility changes over the years. It seems clear that distinct racial and ethnic standards may lead to different values, on average, of joint movement mobility, even though this issue, due to methodological and operational difficulties, has not yet been adequately addressed (Wright 1982).

It also seems that flexibility is genetically influenced, as is clearly evident in some types of hypermobility clinical presentations and in the familial proclivity for some sports (synchronized swimming) and physical performances (circus contortion). Biological cycles, such as the circadian, menstrual, and gestational cycles (Calguneri, Bird, and Wright 1982), cause significant variations in body temperature and levels of some hormones, potentially affecting ligament laxity, muscle and tendon viscoelastic properties, and, therefore, body flexibility. While some research (J.A. Smith 1956) has not found a significant relationship between flexibility (as measured by linear tests) and motor learning, data from our research group (Farinatti, Araújo, and Vanfraechem 1997) suggest that learning to swim is easier for children with wider ankle and shoulder ranges of motion—the movements that are most important for swimming. A more detailed analysis of the influence of various factors on flexibility may be found in another Human Kinetics book, *Science of Flexibility, Second Edition* (Alter 1996).

Flexibility in Health and Disease

Flexibility is a major component of physical fitness; it is important to conduct both simple and complex movements, as well as for sport performance, health maintenance, and daily-activity completion (Cureton 1941; Holland 1968; Harris 1969b; Gersten et al. 1970; Bouchard et al. 1990; Pate et al. 1995; Rejeski, Brawley, and Shumaker 1996; van Heuvelen et al. 1997; Fahey, Insel, and Roth 1999). Flexibility or stretching exercises are nowadays incorporated in almost all exercise programs and have been specifically recommended for both healthy and unhealthy subjects (Pate et al. 1995; Fletcher et al. 1996; ACSM 1998a; Pollock et al. 2000). In contrast to other physical fitness components, however, the relationship between flexibility levels and health is not linear or even direct—low and high extremes may be associated with morbidity and diminished quality of life. In addition, recent longitudinal data have failed to establish a relationship between trunk flexion mobility and risk of mortality in adults (Katzmarzyk and Craig 2002).

Flexibility in Physical Performance

For years, high levels of flexibility have been associated with outstanding performance in sports and dancing. Unlike other variables in the physical fitness domain, such as maximal aerobic power, it is now clear that outstanding levels of flexibility or hypermobility may predispose one to developing abnormal neuromuscular patterns (Russek 1999) and persistent musculoskeletal problems (Loudon, Goist, and Loudon 1998). At the other end of the spectrum, a lack of flexibility and the corresponding possibility of developing chronic back pain or incurring injury related to the practice of sports has also been investigated (Corbin 1984). This section will review the role of flexibility in health, covering aspects related to health maintenance and exercise-related performance.

Daily Activities

Recently, research has been redirected from determining methods to achieve outstanding levels of body flexibility to evaluating the relevance of regularly performed stretching exercises. Regrettably, there are no evidence-based data to indicate what is the ideal level of flexibility for a nonathlete adult. Recent investigations have examined the importance of the physiological range of motion in everyday actions—such as walking (Escalante, Lichtenstein, and Hazuda 2001), standing from a seated position, and even reaching up for an object on a shelf—to primarily reflect the autonomy and independence of an individual (Rikli and Jones 1997). Analysis of the role of flexibility—both the static and dynamic components—in activities of daily living should be multidisciplinary because it relates to different fields, such as morphology, physiology, fitness and wellness, production engineering, biomechanics, and ergonomics. Limitation of joint motion, which occurs over the years as part of the aging process, may impose major restrictions in performing some movements or even prevent them from being made. In an interesting study published over a decade ago in the *Journal of Biomechanics*, Fleckenstein, Kirby, and MacLeod (1988) noted that when knee flexion was limited to 75°, standing from a seated position would only take place if there was a concomitant movement of swinging the arms and flexing the trunk. And yet the peak of the movement, measured in newtons, was almost double that of the movement performed by a subject without knee-flexion restriction, which could seriously compromise a joint previously injured or having a prosthesis (Fleckenstein, Kirby, and MacLeod 1988).

Most activities of daily living require combined and complex movements that involve the coordinated contraction of a number of muscular groups and the combined movement of many joints (Knudson, Magnusson, and McHugh 2000). These actions can be performed by engaging in a number of different combinations of angles and joint movements. For instance, to reach an object that is in front of us and at a height slightly below our hips, we shift the angles of the lumbar spine, hips, knees, and ankles; however, the amplitude of arc for each of these movements may substantially vary from one individual to another, and even according to gender (Thomas, Corcos, and Hasan 1998).

Common daily activities in aged persons, such as walking and reaching, may also be hampered by a decreased range of motion in extremity joints (Escalante et al. 1999; Escalante, Lichtenstein, and Hazuda 1999, 2001). Escalante, Lichtenstein, and Hazuda (2001) observed that about 6% of the variance in walking velocity may be explained by differences in hip and knee flexion. Recently, Brach and VanSwearingen (2002) studied 83 elderly individuals and found that the ability to put on and take off a jacket is strongly related to shoulder range of motion and that gait speed and stride length are related to the active ankle range of motion. On the other hand, Barrett and Smerdely (2002) observed no improvement in health-related quality of life, as assessed by a specific questionnaire called SF36, after 10 weeks of twice-weekly stretching training.

Limitation of joint motion, which happens as part of the aging process, may increase the risk of falling (Brach and VanSwearingen 2002), significantly restrict the performance of some movements, and prevent others from being made at all. It is possible that, in the near future, research will focus on the qualitative and quantitative roles of joint mobility in performing the most important daily-living actions.

Sports and Dance

The idea that high flexibility levels are important for superior performance in sports may be incorrect, as evidenced by the limited and controversial information available (Holland 1968; Travers and Evans 1976; Corbin 1984; Cureton 1941; Leighton 1957a, 1957b; Brodie, Bird, and Wright 1982; Oberg et al. 1984; Lee et al. 1989; Chandler et al. 1990; Craib et al. 1996; Decoster et al. 1997; Araújo 1999b; Nelson et al. 2001a; Jones 2002).

Since Cureton's (1941) and Leighton's (1957a, 1957b) preliminary studies, limited data have been available to effectively determine the role of flexibility in an elite athlete's performance. The scarcity of data in this field may be due to a number of reasons, including the limited interest of athletes and coaches and the lack of standardized flexibility assessment methods according to age groups and level of ability in sports. Another complicating factor is that the expressions *athlete* and *competitive sports* have, in reality, very broad scopes, encompassing a range of sports from archery to gymnastics that vary significantly in terms of performance-related physical fitness profiles. For example, long-distance runners, triathletes, and swimmers rely primarily on their maximal aerobic power to achieve excellent performances, whereas weightlifters and wrestlers require superior muscle strength and power and ice skaters and synchronized swimmers need extraordinarily high flexibility levels to perform their routines.

Specific joint mobility patterns in athletes are associated with the specific biomechanical and technical characteristics of the motor performance of a given sport, or even the position played (Leighton 1957a, 1957b; Nelson et al. 1983; Oberg et al. 1984; Chandler et al. 1990; Ellenbecker et al. 1996; Hahn et al. 1999; Rozzi et al. 1999). Our experience with assessing flexibility in elite athletes confirms this, and we found that in particular cases or sports, an excellent Olympic performance was achieved with an average or below-average flexibility level for the athlete's age group (Araújo, Pereira, and Farinatti 1998). On the other hand, it is likely that very high flexibility levels are needed in athletes whose sport modality involves the application of subjective criteria to assess grace in performance, such as gymnastics, ice skating, diving, and synchronized swimming.

Of recent interest is the possible existence of an inverse relationship between flexibility of the lower-limb joints, particularly the ankle, and performance in walking and running (Knudson, Magnusson, and McHugh 2000). Gleim, Stachenfeld, and Nicholas (1980) were the first to suggest that low-normal flexibility levels are associated with higher movement economy, quantified as the amount of oxygen uptake for a given walking or running speed. Craib et al. (1996) confirmed these results and noted that smaller ranges of motion in ankle dorsiflexion and hip lateral rotation were moderately correlated with movement economy in a 10-minute run at 250 m/minute, whereas the range of motion in trunk flexion failed to show an association. On the other hand, Jones (2002) confirmed that the magnitude of anterior trunk flexion has a significant correlation ($r = 0.68$) with oxygen uptake for experienced runners running at 16 km/h, suggesting that even joints not directly involved with running may be relatively hypoflexible in runners. Thus the question, Should runners avoid stretching exercises? Nelson et al. (2001b) addressed this uncertainty by submitting 32 recreational runners to a 10-week program of flexibility training and observed that increases in the trunk anterior flexion motion (the only movement tested) did not affect the submaximal running economy. Current knowledge holds that somewhat low mobility levels of lower limbs and perhaps of trunk anterior flexion are beneficial for the movement economy of walking and running and that stretching programs do not impair submaximal energy requirements for these activities. However, although somewhat limited ankle mobility may be advantageous for running economy, it may predispose nulliparous varsity athletes to a higher prevalence of urinary incontinence due to differences in the way that impact forces are transmitted to the pelvic floor during exercise (Nygaard, Glowacki, and Saltzman 1996).

An area of recently renewed interest is the effect of stretching exercises on physical perfor-

mance. It has been shown that an intense session of passive stretching of the plantar flexors can substantially decrease isometric strength for over one hour. This type of exercise does not specifically increase muscle protein synthesis and, therefore, is not a stimulus for the muscle hypertrophy that is associated with resistance training (Fowles, Sale, and MacDougall 2000; Fowles et al. 2000; Nelson and Kokkonen 2001; Nelson et al. 2001b). In addition, Church et al. (2001) found that a specific flexibility exercise routine was correlated with a reduction in performance in the vertical jump test results of 40 young women. Investigations are just beginning to explore the mechanisms for the short-lived deleterious effects of flexibility exercises undertaken immediately prior to physical performance (Cornwell, Nelson, and Sidaway 2002). These results suggest that in certain circumstances, flexibility exercises might impair the physical performance of athletes.

Similar to athletes in sports, dancers may also have an optimal flexibility profile that allows for better performances and a decreased chance of injury. Dancers in general must be graceful when they perform, which demands moderate to high static and dynamic flexibility levels. When Grahame and Jenkins (1972) compared the flexibility of 53 London Royal Ballet School students to that of nursing students, they confirmed that the ballet dancers' mobility was significantly higher, especially in trunk anterior flexion. These results were later duplicated in a South African dance school by Klemp and Learmonth (1984), who verified that if they excluded data for trunk flexion—the maximum range of motion of which one can improve with training—differences between the control and ballet-dancer groups would practically disappear. In a follow-up study, Klemp and Chalton (1989) found that after a four-year ballet training period, most of the dancers became hypermobile at trunk flexion, strengthening the association between specific training and acquired trunk flexion hypermobility, although they did not find a relationship between dance performance excellence and generalized hypermobility. In addition to increased trunk flexion, a number of classical ballet positions require wide ranges of motion, particularly in the hip joint (Reid et al. 1987; Gilbert, Gross, and Klug 1998). Typical body flexibility profiles—in particular, a higher hip lateral rotation—exist for both ballet students and more experienced dancers (Khan et al. 1997; Khan et al. 2000; Bennell et al. 2001).

Currently available data indicate that dancers do not exhibit overall hyperflexibility, but they are likely to present higher ranges of localized motion in a few movements instrumental to their performance. Studies on the flexibility of dancers and athletes of differing proficiency levels are needed to broaden our knowledge and yield more definite conclusions and practical implications.

Flexibility in Disease

We have reviewed the ways in which flexibility is related to health and exercise performance, and this section will explore the role of flexibility extremes in unhealthy conditions. A large amount of scientific knowledge on clinical aspects of joint motion has been accumulated in the last decades (Dunham 1949; Carter and Sweetnam 1960; Beighton and Hóran 1970; Rosenbloom et al. 1981; Pitcher and Grahame 1982; American Academy of Pediatrics 1984; Campbell et al. 1985; Handler et al. 1985; Bridges, Smith, and Reid 1992; Westling and Mattiason 1992; Gedalia et al.1993; Grahame 1999; Jessee, Owen, and Sagar 1980; Lewkonia and Ansell 1983; Mikkelsson, Salminen, and Kautiainen 1996; Noyes et al. 1980). Both low and high levels of overall body flexibility, which sometimes represent simple variations from normalcy (Jessee, Owen, and Sagar 1980), have been associated with specific diseases or clinical conditions (American Academy of Pediatrics 1984; Beighton and Hóran 1969; Campbell et al. 1985; Handler et al. 1985; Lewkonia and Ansell 1983; Westling and Mattiason 1992), and flexibility testing may play

a role in the clinical examination of these individuals (Bird, Brodie, and Wright 1979; Bridges, Smith, and Reid 1992; Carter and Sweetnam 1960; Dunham 1949; Noyes et al. 1980; Rosenbloom et al. 1981). These extremes of flexibility have been named *hypermobility* and *hypomobility,* and they will be discussed separately in the following sections.

> Extreme ranges of motion are denominated *hypomobility* and *hypermobility*, respectively, for small and large arcs of movement. While these terms are more often used to describe overall flexibility, they can also be used for specific joint movements.

Hypermobility

The phenomenon of hypermobility and its hereditary nature were perceived long ago, as evidenced in particular paintings of Rubens, Grünewald, and Traversi from the 15th and 18th centuries (Dequeker 2001). Key (1927) reported that hypermobility is hereditary in nature and is characterized by extreme joint mobility (figure 2.1). He also noted that most individuals with hypermobility should not be considered abnormal, but rather the bearers of uncommon anatomical features. Since then, the use of the term *flexibility* has been quite

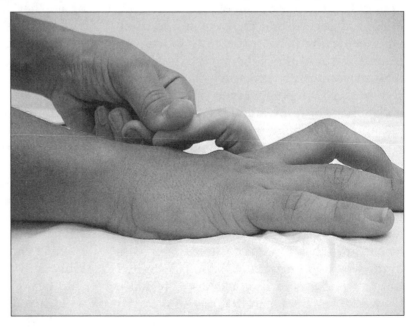

Figure 2.1 Hypermobility.

common in the medical field, particularly in rheumatology and related areas.

The term *hypermobility syndrome* was probably used for the first time by Dr. J.A. Kirk and colleagues in June 1967 in an article published in the *Annals of the Rheumatic Diseases* (Kirk, Ansell, and Bywaters 1967). Later, many authors used it in a medical sense (Wood 1971; Wynne-Davies 1971; Scharf and Nahir 1982; Biro, Gewanter, and Baum 1983), and some included the term *benign* (Jessee, Owen, and Sagar 1980; Grahame and Bird 2001). The so-called Berlin nosology (Grahame 1999; Russek 1999) has proposed the term *familial joint hypermobility syndrome* to highlight the genetic component, but its use is not widespread.

Although no consensus exists on the medical criteria diagnostic for hypermobility, Klemp's approach (1997) seems valid (see discussion of Beighton-Hóran criteria in chapter 3, page 43). Abnormally high serum levels of growth hormone, insulin, and IGF-1 are present in adults with hypermobility syndrome, and if confirmed, their measurement may be an interesting possibility for establishing a laboratory diagnosis (Denko and Boja 2001). While the presence of hypermobility is typically benign and the prognosis is therefore quite favorable, it is often the only external sign of complex medical conditions, particularly Ehlers-Danlos and Marfan syndromes (Grahame 2000c). Marfan syndrome is frequently associated with sudden death during physical exercise (Maron et al. 1996), a fact that enhances the importance of establishing a differential diagnosis.

The prevalence of hypermobility depends on the criteria used to characterize it. Despite methodological differences in different studies, hypermobility is a fairly common finding, ranging from 1 to 35% of the population (Forleo et al. 1993; Decoster et al. 1997; Grahame 1999; Russek 1999; Seow, Chiow, and Khong 1999; Duró and Vega 2000). Prevalences of hypermobility and ligamentous hyperlaxity display hereditary (Sturkie 1941; Carter and Sweetnam 1960; Wynne-Davies 1970) and polygenic (Grahame 1999) components, most often dominant (Beighton and Hóran 1970; Russek 1999; Martin and Ives 2002), but sometimes recessive (Hóran and Beighton 1973). There are clear ethnic dif-

ferences (Beighton, Solomon, and Soskolne 1973; Wordsworth et al. 1987; Forleo et al. 1993; Birrell et al. 1994; Mikkelsson, Salminen, and Kautiainen 1996; Rikken-Bultman et al. 1997; El-Garf, Mahmoud, and Mahgoub 1998; Seow, Chiow, and Khong 1999; Verhoeven, Tuinman, and Van Dongen 1999; Vougiouka, Moustaki, and Tsanaktsi 2000), and prevalences tend to be higher in women and to decrease with age in adults (Beighton, Solomon, and Soskolne 1973; Dungy and Leupp 1984; Larsson, Baum, and Mudholkar 1987; Mikkelsson, Salminen, and Kautiainen 1996; Decoster et al. 1997; El-Garf, Mahmoud, and Mahgoub 1998; Qvindesland and Jónsson 1999), after reaching their peaks at age two to three (Wynne-Davies 1970). Hypermobility can also be seen in only one joint movement, e.g., *genu recurvatum* (Loudon, Goist, and Loudon 1998), in a few joints (referred to as *pauci-articular* when in less than five joints), or to be of a more general nature (Grahame 1999).

Considering the relatively high prevalence of joint hypermobility and the medical interest in establishing relationships between signs and syndromes, it is no wonder that in a number of studies hypermobility has been related to other medical features or situations. A number of papers were published from the late 1960s to the early 1980s on the subject. However, probably because of difficulties in establishing an etiological diagnosis and providing proper treatment for hypermobile individuals, interest waned. More recently, attention to the issue has increased as other associations and medical implications have been identified. Despite the renewed interest, this medical condition is still relatively unknown and neglected, even by rheumatologists (Grahame and Bird 2001), and particularly by those who work in the fields of orthopedics, physical medicine, and physical therapy (Russek 1999). This causes much suffering and distress in people who experience joint hypermobility (Grahame 2000b), many of whom are considered hypochondriacs (Russek 1999). For a detailed discussion of this topic, consult the excellent review articles recently published by Drs. Rodney Grahame and Patrick Klemp (Klemp 1997; Grahame 1999, 2000a, 2001).

Sutro (1947) described five cases of extreme joint mobility and suggested that ligamentous and capsular tissue hyperextensibility are responsible for hypermobility. This conclusion is still highly regarded, because it is believed that benign joint hypermobility syndrome is caused by excessive ligamentous laxity associated with changes in the genes responsible for the synthesis of three proteins, collagen, elastin, and fibrillin (Grahame 1999). *Benign hypermobility*, as the first of a broad range of changes that takes place in the connective tissue, is a form fruste of a genetically acquired dysfunction (Grahame 1999, 2001). Medically, the superposition of features within this range is large, so the distinction between benign hypermobility syndrome and type III Ehlers-Danlos syndrome is minimal, if any. Hypermobility is rarely seen in individuals with type IV Ehlers-Danlos, which is called the vascular form because of the heightened risk of rupturing vascular and visceral vessels and is characterized by a mutation of the type III protocollagen gene (Pepin et al. 2000). On the other hand, hypermobility is a prominent feature in some other types of Ehlers-Danlos syndrome.

No difference in the passive absorption of energy for a given submaximal joint movement angle has been found in women with and without overall hypermobility (Magnusson et al. 2001). These data suggest that passive properties of a muscle-tendon unit are not different across conditions and that the higher flexibility of hypermobile women is not caused by changes in the physiological properties of the muscle-tendon unit (Magnusson et al. 2001).

Investigations into the genetic basis of benign joint hypermobility syndrome have been done, but the loci and mutations that cause ligamentous hyperlaxity have not been precisely defined (Grahame 1999). The syndrome is quite likely polygenic and affects the structures of types I and III collagen molecules, with a relatively higher proportion of type III (Russek 1999; Grahame 2000c). Recent data suggest that an anomaly in the q24-26 region of chromosome 15 may be related to joint hyperlaxity and the psychiatric disorders often exhibited by these individuals (Gratacos et al. 2001).

Hypermobility can cause a number of unspecific symptoms, leading to psychosocial problems and often significantly affecting the quality of life of those who present with it (Grahame 2000c). Hypermobility eventually leads to arthralgia, back pain, and frequent and recurrent torsion or luxation (Nef and Gerber 1998). This section of chapter 2 addresses the principal clinical aspects of joint hypermobility—which is most frequently determined by ligamentous hyperlaxity tests—including its association with musculoskeletal changes and mitral valve prolapse.

Musculoskeletal Disorders

Hypermobility often presents a range of histological and physiological features, including changes in the composition and physical properties of connective tissue proteins and some undetermined neurophysiological influences, such as reduced joint proprioceptive acuity, nociceptive enhancement, and greater tendency toward depression (Grahame 2001). These features, noted particularly in children and adolescents, have spurred many authors to study the association between joint hypermobility and musculoskeletal and joint symptoms (Carter and Sweetnam 1958; Kirk, Ansell, and Bywaters 1967; Finsterbush and Pogrund 1982; De Inocencio 1998; Hudson et al. 1998; Mikkelsson et al. 1998; Goldman 2001).

When Kirk, Ansell, and Bywaters (1967) reported the existence of a relationship between musculoskeletal complaints and the presence of overall hypermobility, they established the basis for hypermobility syndrome. Finsterbush and Pogrund (1982) observed that among hypermobile individuals, 42% reported some mild discomfort or diffuse musculoskeletal symptoms. Most had multijoint complaints, and these more often involved the feet and knees. Even more interesting is that only 10% of the individuals denied any musculoskeletal symptoms, instead presenting only with scoliosis. Acasuso-Diaz, Collantes-Esteves and Sanchez-Guijo (1993), investigating young soldiers, showed that those with higher ligamentous laxity were more likely to incur ankle injury over two months of military training.

Children and adolescents frequently report in joints and muscles low- or average-intensity pain that is generally difficult to characterize, is intermittent, may be localized or more diffuse, and quite often tends to migrate throughout the body (Mikkelsson et al. 1998). It is necessary to clinically differentiate a rheumatic or chronic arthritis condition from a condition in which hypermobility signs predominate (Scharf and Nahir 1982; Lewkonia and Ansell 1983; Gedalia et al. 1985; Grahame 1999). It is seldom possible to establish a clinical diagnosis of Ehlers-Danlos syndrome when extreme hyperlaxity manifestations, particularly in the hand and wrist joints, are presented with hyperextensibility abnormalities and skin frailty (Kornberg and Aulicino 1985). While some authors suggest that hypermobility has an etiological role in musculoskeletal pain in children and adolescents (Gedalia et al. 1993), this has not been systematically seen in larger samples. Other authors believe these complaints are caused, at least in part, by psychosomatic disorders (Mikkelsson et al. 1998) and by direct trauma or injury due to excessive use (De Inocencio 1998).

Hypermobility has also been associated with fibromyalgia. Hudson et al. (1998) confirmed that general hypermobility is more prevalent in patients with rheumatic conditions of the soft tissues, including fibromyalgia. Corroborating these findings, Acasuso-Diaz and Collantes-Estevez (1998) investigated 133 adult Spanish females and found that the prevalence of hypermobility was at least twice as high among those with diagnostic criteria for fibromyalgia, suggesting that ligamentous hyperlaxity could have an important role in the pathogenesis of pain in fibromyalgia. On the other hand, Karaaslan, Haznedaroglu, and Ozturk (2000) questioned these findings and suggested that hypermobility was not effectively greater in patients with definite criteria for fibromyalgia, but only in patients who had partial criteria for fibromyalgia and presented with overall muscular pain. It is possible that contradictions among the different studies are due to selection bias in the investigated populations, and as yet there are not any prospective and randomized studies to definitely answer these questions.

> Ligamentous hyperlaxity is identified by the excessive mobility of joints in which tendons and ligaments play a significant role in limiting range of motion.

Al-Rawi and Nessan (1997) demonstrated that patellar chondromalacia was almost four times more prevalent in patients, especially women, with knee hypermobility, suggesting an etiological role for this disease. There are also reports that osteoarthritis and chondrocalcinosis may be late complications in extremely mobile joints (Bird, Tribe, and Bacon 1978). More recently, Punzi et al. (2001) reported that hypermobility was a common sign in their rheumatic arthritic female patients. In addition, ligamentous hyperlaxity has often been found in individuals with idiopathic scoliosis (Veliskasis 1973) and in those, particularly the women, who seek specialized clinics for correcting limb congenital defects (Hassoon and Kulkarni 2002).

There seems to be a moderate association between overall and localized joint hypermobility and the occurrence of musculoskeletal or joint clinical conditions (Wynne-Davies 1970). Differential diagnosis, including lab data, should be done in the presence of joint hypermobility and ligamentous hyperlaxity to rule out the possibility of rheumatic disorders. Early identification of joint hypermobility may contribute to a better therapeutic approach and proper exercise prescription (Grahame 2001). Thus, it seems appropriate to include well-standardized joint motion tests in the clinical assessment of children and adults with unspecific musculoskeletal complaints, particularly women and those in whom one cannot easily identify an etiological reason.

Mitral Valve Prolapse and Other Cardiovascular Disorders

Individuals with joint hypermobility may present with changes in connective tissue not only in the structures directly involved with locomotion, but also in other body systems. It therefore seems appropriate to review links between joint hypermobility and abnormal structures of the cardiovascular system.

Among structural changes of the heart, the most common is certainly prolapse of the mitral valve, with a prevalence of 1 to 5% of the adult population. Mitral valve prolapse, which is more frequent in women, may be defined as an anomalous protrusion of one or two mitral valve leaflets into the atrium during ventricular systole. It occurs because of exaggerated elasticity of the leaflets or one of their tendon components and has a clear genetic component. In most people, there is neither reflux nor significant symptomology, and clinical prognosis is excellent.

Because changes in collagen are features of both medical conditions (Malcolm 1985), it is quite reasonable to believe that individuals with mitral prolapse typically would be hyperflexible. The population of ballet dancers, in which there is a high prevalence of joint hypermobility and ligamentous hyperlaxity (Grahame and Jenkins 1972), is a convenient model for studying the relationship between joint hypermobility and mitral valve prolapse. Although Klemp and Learmonth (1984) found no cases of mitral valve prolapse among a group of male and female ballet dancers, Cohen et al. (1987) found a very high prevalence of 42% in a population of 44 female professional ballet dancers. This difference may be explained by Klemp and Learmonth's inclusion of males, who are less likely to have the valve disorder, and by their group's relatively low prevalence of hypermobility (less than 10%). Different criteria may also have been used to identify the presence of mitral valve prolapse.

Marks et al. (1983) did not find hypermobility among male or female adults with mitral valve prolapse, but many other researchers have found important associations between the two clinical conditions (Grahame et al. 1981; Pitcher and Grahame 1982; Handler et al. 1985; Rajapaske et al. 1987; Ondrasik et al. 1988; Rodriguez et al. 1991; Bulbena et al. 1993). The incidence of mitral valve prolapse in hyperflexible individuals varies from 2 to 10 times higher than in individuals of the same age without hypermobility, reaching up to 30% of the sample (Grahame et al. 1981; Russek 1999). Many of these individuals also present with musculoskeletal symptoms that would go unnoticed in a strictly cardiological assessment (Ondrasik et al. 1988). In our experience (Chaves, Araújo, and Araújo 2001), the presence of ligamentous hyperlaxity signs was four to five times more common in women with mitral valve prolapse, who tend to be more flexible in most of the principal joint movements.

A more life-threatening cardiovascular disorder in which mitral valve prolapse is prominent and common is Marfan syndrome. In this syndrome, which is an autosomal dominant disorder of the connective tissue, there is mutation of the fibrillin-1 gene encoding for fibrillin, a major component of the extracellular microfibrils. Its prevalence is about 1 in 10,000 individuals. Individuals who have Marfan syndrome are at high risk for aortic dissection or rupture, which is often fatal and contributes to the low mean life expectancy of 32 years in untreated Marfan cases (Nienaber and Von Kodolitsch 1999). In this syndrome, hypermobility has an ancillary role. Grahame and Pyeritz (1995) investigated 27 children and 48 adults having an established diagnosis of Marfan syndrome and confirmed the existence of joint hyperextensibility in about 85% of the cases. Hypermobility prevalence gradually diminished with age. Marfan syndrome cases in which extreme hypermobility is the main clinical feature are quite rare (Walker, Beighton, and Murdoch 1969).

These studies show the importance of using a multidisciplinary approach to assess the relationship between joint hypermobility and cardiovascular structural changes. For instance, in the presence or suspicion of mitral valve prolapse and its typical symptoms—tachycardia, palpitations, anxiety, and breathlessness—it may be clinically relevant to apply joint mobility and ligamentous hyperlaxity measurement tests.

Other Hypermobile States

A number of other associations have been made—some even by accident—between joint hypermobility and specific clinical conditions. While it is sometimes possible to establish relationships in these entities with changes in the connective tissue, in others the mechanisms are unknown and it is difficult, or even may be impossible, to establish cause-and-effect relationships.

It seems that some chromosomal alterations predispose one to ligamentous hyperlaxity and joint hypermobility. For instance, Baughman et al. (1974) confirmed that the carrying angle (or cubital angle) reached the maximum with the XO phenotype and the minimum with extranumerary chromosomes X or Y. According to Biro, Gewanter, and Baum (1983), people with chromosome 21 trisomy (Down's syndrome) may also present with hypermobility. Children with Down's syndrome tend to be more flexible than healthy peers of the same age (Semine et al. 1978; Parker and James 1985), even though cases of extreme ligamentous hyperlaxity are infrequent (Livingstone and Hirst 1986). In addition, these individuals seem to have instability in the cervical spine, particularly in the atlantoaxial joint (Semine et al. 1978), which tends to be asymptomatic. This problem led the American Academy of Pediatrics Committee on Sports Medicine (1984) to recommend that every child or adolescent with Down's syndrome who engages in sports that carry a risk of head or neck trauma or injury periodically submit to clinical and radiological assessment before being granted permission to practice those sports.

Hypermobility—specifically, changes in the position of the hip and relaxation of the pelvic ligaments—is common during pregnancy (Abramson, Roberts, and Wilson 1934). These changes are caused by the hormone relaxin, the serum level of which increases ten-fold over the last four weeks of pregnancy (Calguneri, Bird, and Wright 1982). These authors found higher extensibility of the metacarpophalangeal joint during the second half of pregnancy compared with that during the first weeks postdelivery (Calguneri, Bird, and Wright 1982), similar to the findings of Dumas and Reid (1997) on knee-ligament laxity. Also related to pregnancy, a recent prospective pilot study (Tincello, Adams, and Richmond 2002) tested if joint motion measurements made in nulliparous pregnant women could predict the incidence of postdelivery urinary incontinence. The authors observed that although scores for overall hypermobility were not useful for this purpose,

assessment of the presence or absence of elbow hyperextension had 75 to 80% sensitivity and specificity and 14% positive and 99% negative predictive values. According to the authors' conclusion, it is still too early to judge the medical merit of routine elbow mobility assessment in prenatal tests; however, this is currently a very attractive possibility.

For more than 20 years, the al-Rawis have been studying women with uterine or vaginal prolapse, half of whom had delivered at home and had not had postnatal rest. The prevalence of overall hypermobility was at least three times higher in those women than in members of the control group, who did not have genital prolapse (al-Rawi and al-Rawi 1982). Norton et al. (1995) indicated that the different forms of genitourinary prolapse were at least two times more prevalent in women with overall joint hypermobility. Analysis by McIntosh et al. (1996) of the angular measurements of flexibility in various joints of women with different types of Ehlers-Danlos syndrome showed that hypermobility was correlated to prolapse in the pelvic floor. The study also revealed a significant and direct association between measurements of wrist joint range of motion and the presence of urinary incontinence. It is unknown, however, if these results apply to a population with a normal prevalence of joint hypermobility.

Another area of interest is the relationship between hypermobility and psychiatric symptoms or conditions. Bulbena et al. (1993) found that phobias and panic syndrome were very likely to occur in hyperflexible patients, at an odds ratio of 10:7; this is a much higher ratio than that expected for the concurrence of anxiety and mitral valve prolapse. In a more recent study by the same group (Martin-Santos et al. 1998), the prevalence of hypermobility syndrome in individuals with anxiety-related disorders was 67%. They also provided evidence that patients with medically relevant anxiety had a 16-fold higher likelihood of being hypermobile than did nonanxious individuals (Martin-Santos et al. 1998), which suggested the existence of a constitutional predisposition for these medical conditions to be associated with each other. Recent data suggest that mutations of chromosome 15q may be associated with panic, anxiety, and joint hypermobility conditions (Ewald, Rosenberg, and Mors 2001).

New associations between hypermobility and clinical conditions have periodically come to light in the literature of the different fields of medicine and health sciences. Kaplinsky et al. (1998) found that hyperflexibility of the thumb may be associated with a mild tendency to bleed. De Felice et al. (2001) identified a correlation between joint mobility and a history of pyloric hypertrophic stenosis in childhood. In addition, Hall et al. (1995) found that hypermobile subjects have poor proprioceptive feedback, which may be related to their increased risk of knee overextension lesions.

It is likely that the growing interest in the field of flexibility, particularly in measuring and assessing joint mobility, may lead other research groups to report new, interesting, and medically relevant associations with joint hypermobility.

Hypomobility

In opposition to the hypermobility syndromes described in the previous section are clinical conditions in which limited joint motion represents a major finding (Rosenbloom et al. 1981; Campbell et al. 1985; Len et al. 1999). It should be pointed out that hypomobility (figure 2.2) as mentioned in this text differs conceptually from physiological decrements typically seen with the aging process.

Hypomobility has been seen in different types of pathological states, such as low-back pain (Bach et al. 1985; Lankhorst, Van De Stadt, and Van Der Korst 1985; Ellison, Rose, and Sahrmann 1990; Kujala et al. 1997; Harreby et al. 1999), spondylolisthesis (Phelps and Dickson 1961), poststroke sequelae (Bressel and McNair 2002), diabetes mellitus (Kennedy et al. 1982; Rosenbloom et al. 1983; Rosenbloom et al. 1984; Campbell et al. 1985; Pal et al. 1987; Arkkila, Kantola, and Viikari 1997), and diabetes insipidus (Fitzgerald, Greally, and Drury 1978). This section will review the major clinical conditions in which hypomobility is a relevant feature.

Low-Back Pain

Low-back pain occurrence is extremely common in an adult's life, and when asked, most people will report experiencing at least one episode. Since early-life onset has been related to chronic low-back pain during adult life, Feldman et al. (2001) studied a large sample of Canadian adolescents and found that tight hamstring tendons represent a risk factor for development of back pain in following years. Although most cases of low-back pain are self-limited and of light to mild intensity, in some cases pain can be severe enough to produce significant physical disability.

The attempt to relate low-back pain, a major public health issue, with hypomobility of the trunk and hamstrings is quite old, and in reality, many of the flexibility testing protocols are based on this premise. Nevertheless, there remains some controversy over this issue (Kirby et al. 1984; Bach et al. 1985; Ellison, Rose, and Sahrmann 1990). Theoretically, people with restricted arcs of movement in the low trunk and hamstring regions could be more prone to unduly compressing or hurting nerve roots coming off the spine during sudden or intensive movements. On the other hand, an

Figure 2.2 Hypomobility.

abnormally large range of motion in the lumbar region could also increase the propensity for similar damage.

While this relationship is far from being elucidated, some studies have provided interesting data. Ellison, Rose, and Sahrmann (1990) reported that asymmetry between medial and lateral hip rotation may be more important etiologically than true hypomobility. Of over 150 individuals, those who exhibited a greater lateral hip rotation range of motion than a medial range were more prone to present with low-back dysfunction. Kujala et al. (1997) performed a longitudinal follow-up of lumbar mobility and low-back pain epidemiology during adolescence and found that those in the lower tertile of maximal lumbar extension had a three to four times higher chance of developing low-back complaints in a three-year period than did those in higher tertiles. These data apparently contradict previous results (Kirby et al. 1984) reporting that female gymnasts with low-back discomfort had greater toe-touching ability than those without symptoms.

After reviewing the literature, it seems that the relationship between flexibility and low-back pain could have a biphasic or a U-type pattern, in which both hypomobility and hypermobility can increase low-back pain occurrence. This may explain why both sedentary individuals and gymnasts—most likely by distinct mechanisms—can present with similar symptoms. It is even possible that lumbar spine mobility is not an issue in low-back pain, according to recent data from a study by Nattrass et al. (1999). If this interpretation holds true, it may also explain why a large study of almost 3,000 adults (Jackson et al. 1998) with flexibility levels ranging from low to high recently found that classical flexibility testing—sit-and-reach—did not predict low-back pain, since plotting of biphasic data provides a very low correlation coefficient.

In summary, hypomobility and low-back pain are often linked. Although this finding is accurate in many cases, the linkage may be etiologically inappropriate because low-back pain has also been associated with hypermobility (Sutro 1947). An interesting interventional study (Lankhorst, Van De Stadt, and Van Der Korst 1985) followed 31 patients with idiopathic low-back pain whose symptomatology improved when their lumbar mobility was reduced, corroborating that, at least for these patients, hypermobility rather than hypomobility was associated with low-back pain.

Diabetes Mellitus

Clinical hypomobility, specifically in the hands, has been seen in diabetic patients (Lundbzaek 1957). However, it was only in 1974 that pediatric endocrinologists in Florida presented a new syndrome based on three cases that were characterized by limited joint motion and growth deficiency in childhood diabetes mellitus (Rosenbloom and Frias 1974). This preliminary report was followed by other studies from the same research group that incorporated a larger number of patients and other variables (Grgic et al. 1975; Grgic et al. 1976; Rosenbloom et al. 1981; Rosenbloom et al. 1982; Rosenbloom et al. 1983; Rosenbloom et al. 1984). They were able to show that diabetic children with limited joint motion had normal serum growth hormone levels (Grgic et al. 1975) and that there was a relationship between the duration of the disease and the prevalence and significance of joint changes (Grgic et al. 1976). In a large sample, they found joint motion was limited in 28% of children with diabetes compared to barely 1% in healthy children (Grgic et al. 1976).

Diabetic children with limited joint motion have a significantly higher risk for the development of long-term microvascular complications (Rosenbloom et al. 1981). Other relevant clinical findings identified include the ranking of 75% of diabetic children with limited joint mobility in the lowest 25th percentile for height, a three-times-higher incidence of retinopathy, and a twofold increase in symptomatic neuropathy (Rosenbloom et al. 1982; Rosenbloom et al. 1984; Starkman et al. 1986). These findings have been confirmed by other research groups (Benedetti et al. 1975; Benedetti and Noacco 1976; Starkman and Brink 1982; Fitzcharles et al. 1984; Kennedy et al. 1982; Rossi and Fossaluzza 1985; Madácsy et al. 1986; Starkman et al. 1986).

Limited joint motion is a soft-tissue consequence of diabetes mellitus. It is mainly restricted to the hands, starting in the fifth finger and moving slowly to other fingers (Kennedy et al. 1982; Rosenbloom et al. 1982; Shinabaerger 1987). Patients seldom spontaneously report it, because it does not cause any functional impairment (Rosenbloom et al. 1982; Fitzcharles et al. 1984). This clinical sign's high incidence—somewhere between 8 and 36%—and elevated specificity make it useful for screening and prognosis (Kennedy et al. 1982; Rosenbloom et al. 1983; Campbell et al. 1985). The first symptom is stiffness in the hands, which normally appears around the second decade of life and is followed about two years later by microvascular angiopathies and involvement of other, larger joints (Rosenbloom et al. 1984; Shinabaerger 1987). If hypomobility starts earlier in life, before puberty's growth spurt, the final height achieved is lower (Rosenbloom et al. 1982). In an interesting report, chest physicians associated with Rosenbloom (Schnapf et al. 1984) identified reduced lung compliance and increased elastic recoil of the lungs, which induce a restrictive respiratory pattern, in diabetic individuals.

More recently, Arkkila, Kantola, and Viikari (1997) found that, when confounding factors such as age and duration of diabetes were controlled, diabetic patients who are hypomobile had a three- to four-times-higher risk of developing coronary and cerebrovascular diseases or nephropathy and an even higher risk of proliferative retinopathy. In another study, limited active and passive ankle motion was found to be associated with a cutaneous sensory deficit (Simmons, Richardson, and Deutsch 1997), which has practical implications for foot care and physical activity guidelines for diabetic patients.

While the biochemical basis for diminished mobility still is not fully understood, it is clear that it includes collagen changes—mainly an increase in the number of molecular interconnections or cross-linkages and higher levels of glycosylation (Campbell et al. 1985). Despite Grgic et al.'s (1976) suggestion many years ago that it is an interesting area of investigation, so far, the fifth finger's passive resistance and dynamic flexibility have not been formally evaluated in diabetic patients. Doing so could provide relevant clinical information.

Other Hypomobile States

In addition to diabetes mellitus and low-back pain, other clinical entities also present hypomobility as a prominent feature that may affect daily life. The ambulatory techniques of people with chronic arthritis differ significantly from those of healthy subjects in terms of rate and range of knee motion (Brinkmann and Perry 1985). In juvenile rheumatic arthritis, the substantial loss of range of motion in affected joints provokes functional disability (Len et al. 1999). Recent data (Cranney et al. 1999) have shown that the limited joint motion usually seen in this disease is associated with genetic and inflammatory processes, as demonstrated by specific markers. In a rare syndrome combining two forms of diabetes—mellitus and insipidus—and optic atrophy, hypomobility is also found; more specifically, a fixed flexion deformity of all interphalangeal joints was found in two siblings (Fitzgerald, Greally, and Drury 1978).

Hemophilia, a condition in which the blood does not clot, is also associated with hypomobility. Johnson and Babbitt (1985) described degenerative joint changes, including hypomobility, in 48 hemophilic patients who also presented with functional limitations and radiological alterations.

With the development and widespread use of flexibility testing in clinical settings, it is likely that in the future other diseases will reveal partial or general hypomobility to be part of their signs and symptoms.

Flexibility in Injuries and Delayed Muscle Soreness

The practice of sports is frequently associated with locomotor system injuries. These injuries may occur with direct trauma, as in a collision between two players, or be due to an indirect mechanism, such as ligamentous rupture caused by an abnormal movement of the joint. Injuries also depend on extrinsic factors, such as the type of sport, how it is practiced, environmental and material conditions, and intrinsic features, such as the physical, psychological, and social condition of the individual (Lysens et al. 1984). Frequently, injuries are due to the overuse of a certain structure of the locomotor system. For example, McHugh et al. (1999) found that stiffer subjects were more prone to develop exercise-induced muscle damage after six bouts of medium-speed isokinetic eccentric submaximal hamstring contractions. Generally speaking, these lesions are more common in individuals submitted to long-term, high-intensity physical training programs, such as triathletes or soldiers. One of the main reasons to study flexibility is to assess the role of such variables in preventing and thus lessening the incidence of physical activity-related injuries.

For many years, it was believed that stretching exercises and "adequate" flexibility levels helped to prevent sports lesions. In an excellent review article, Gleim and McHugh (1997) analyzed data from 18 studies, most of them retrospective, and concluded that there is no definite evidence suggesting that performing stretching exercises prevents injuries. More recently, Shrier (1999), expanding the scope of the review to include papers published in French, confirmed that stretching exercises performed before a physical activity do not prevent injuries. The author presented five reasons why flexibility exercises do not prevent injuries:

1. In animals, the heat-induced increase in muscular compliance favors tissue rupture.
2. Pre-workout stretching has no practical effect in activities in which the maximum range of joint motion is not reached, such as an easy run.
3. Stretching does not affect muscular compliance during the eccentric phase, which is when most injuries occur.
4. Stretching itself may damage some structures.
5. Stretching may mask muscular pain in humans.

It is best to have a less dogmatic approach in the real world. If, on one hand, it seems clear that stretching before engaging in activities such as walking, jogging, and cycling will not reduce the already low chance of injuries in nonathletes, it may still be helpful to tennis players, skaters, and gymnasts, whose performance requires precision and short bursts of high-intensity movements. For these groups, it seems wise to have a warm-up routine, including exercises during which the maximum range of joint movement is reached, even if there is not (and may never be) a scientific confirmation of benefit.

Although the acute effect of stretching exercises does not seem to be important for preventing injuries in sports, the role of flexibility in injury prevention is less certain when comparing the incidence of sports injuries in people with differing levels of flexibility. For instance, Kirby et al. (1984) noted that more-flexible female gymnasts had a higher incidence of low-back pain; McMaster, Roberts, and Stoddard (1998) found a correlation between shoulder laxity and shoulder pain in elite swimmers. On the other hand, Jönhagen, Nemeth, and Eriksson, (1994) studying sprinters, and Hartig and Henderson (1999), following recruits undergoing military training, verified that less-flexible individuals presented with higher injury incidences. It seems that other variables are involved, such as the type of sport and gender. In dance students, Wiesler et al. (1996) found no relationship between ankle flexibility and the occurrence of injuries in a one-year period. Tyler et al. (2001) found that there was no difference in hip adductor mobility in professional ice hockey players who had or had not suffered injuries in hip adductor muscles during the season. Finally, Krivickas and Feinberg (1996) investigated 201 university students of both genders and verified in men only an inverse relationship between ligamentous laxity and lower-limb injuries. Based on the literature, it seems appropriate to agree with Hartig and Henderson's (1999) suggestion that the relationship between the incidence of injuries and flexibility is U-shaped, showing that extreme flexibility levels tend to be associated with a higher incidence of injuries. Most recently, Herbert and Gabriel (2002) performed a systematic review of the literature and found no evidence that stretching routines performed before or after exercise protect against delayed muscle soreness.

When discussing the role of flexibility in injury prevention and incidence, it is also important to note that stretching exercises themselves may cause muscle and tendon injuries, especially if the stretching is inadequately performed. For example, when Askling et al. (2002) questioned 97 student-dancers, they found that a large majority had had at least one hamstring injury during a 10-year period, almost 90% of which had occurred during slow stretching preliminary exercises. In individuals not used to this type of training, Smith et al. (1993) noted that a single stretching exercise session, whether static or ballistic, increased serum levels of creatine kinase and produced moderate delayed muscle soreness. Prospective controlled randomized clinical trials are necessary for a better understanding of the relationships between overall and specific flexibility and physical activity-related injuries.

Classical Flexibility Testing

This chapter briefly reviews the many methods and protocols available for testing flexibility. First, major contributions relevant to flexibility testing are discussed in a historical perspective. We then analyze the different classification systems available for flexibility testing, including our own 18-criteria proposal. In the process we present a synopsis of some of the most-used methods for flexibility testing in different professional areas. Finally, instruments and devices used for range of joint motion measurements are discussed.

Overview and Historical Perspectives

Flexibility is widely recognized as one of the main components of physical fitness (Cureton 1941; Corbin and Noble 1980; Reilly 1981; Bouchard et al. 1990; Pate et al. 1995; Borms and Van Roy 1996; Fahey, Insel, and Roth 1999; Cooper Institute for Aerobic Research 1999; ACSM 2000). Therefore, it is logical to integrate it into physical fitness test batteries (Silman, Haskard, and Day 1986; Skinner, Baldini, and Gardner 1990; Borms and Van Roy 1996; Suni et al. 1998; Cooper Institute for Aerobic Research 1999; ACSM 2000). However, it is inappropriate to consider flexibility as a single and uniform variable or characteristic, because it is specific for a given joint movement (Harris 1969). As such, a single arc of joint motion probably will not reflect the individual's overall mobility or suppleness, the consideration of which is mandatory in flexibility testing. It is likely that certain ranges of joint motion are associated with health and performance needs; indeed, extremely low and high levels of range of motion are commonly associated with un-

healthy conditions, as discussed in chapter 2. This characteristic distinguishes flexibility from other health-related physical fitness components (such as maximal aerobic power and maximal muscle strength and power), in which higher values typically relate to higher standards of health and physical performance in all age groups. In this sense, flexibility may more closely resemble body composition, in which a limited range of values is desirable and extreme results are often related to disease or abnormal conditions.

Despite its special characteristics, which should be taken into account in evaluation programs, it is well recognized that flexibility testing can offer precious information in the context of physical fitness assessment. Figure 3.1 presents various potential and practical situations in which flexibility measurement can aid in the development of health and fitness strategies.

That there are a large variety of protocols, techniques, methods, instruments, and devices for flexibility testing attests to the relative importance and application of flexibility in different settings. For example, flexibility testing may be used by physical educators as part of their physical fitness appraisal (Cureton 1941), by team physicians as an instrument for injury risk assessment during a competitive season (Nicholas 1970), by rheumatologists to identify benign hypermobility in a young woman (Biro et al. 1983), and by pediatric endocrinologists to determine an insulin-dependent diabetic adolescent's potential for microangiopathy (Rosenbloom et al. 1981).

Flexibility testing is not a new issue. As a matter of fact, some studies on this topic took place in the 1800s, as discussed by Elward (1939). In the late 19th century, Potter (1895)

Flexibility testing is useful for:

1. Physical fitness assessment
2. Preparticipation medical or functional evaluation for exercise programs
3. Assessment of injury risk potential
4. Baseline data for assessing results of intervention (either physical training or rehabilitation)
5. Diagnosis of causes of poor or limited performance in sport or daily living activities
6. Assessment of outstanding potential for specific sport modalities
7. Clinical diagnosis and follow-up of hypo- or hypermobility conditions

Figure 3.1 Potential and practical uses of flexibility testing.

suggested that the cubital angle, which was measured by a device similar to current goniometers, was larger in female than in male subjects. At the time of the paper's publication, this approach was very helpful for gender identification in medicolegal issues. However, when viewed retrospectively, it is clear that major interest in flexibility testing coincided with the need to assess disability—especially loss of range of motion—in response to two conditions: widespread poliomyelitis at the beginning of the 20th century and injuries suffered by participants in World War I (Albee and Gilliland 1920; Alquier 1916; Gilliland 1921). Albee and Gilliland (1920) first proposed the use of metrotherapy, a treatment strategy in which therapy-induced improvements in range of motion were quantified in injured soldiers, who, interestingly, manufactured the instruments themselves. In the following year, Gilliland (1921) published what are likely the first range of joint motion norms, based on data obtained in 100 healthy male adults.

Around 1940, there was a surge of published studies on flexibility testing. Both active and passive ranges of motion in pronation and supination were statistically compared (Glanville and Kreezer 1937), norms for ranges of motion in different joints were proposed by distinct groups (Wiechec and Krusen 1939; Fisk 1944; West 1945), and the first large study of children's flexibility was conducted (Gurewitsch and O'Neill 1941). During this time, Cureton (1941) thoroughly discussed flexibility as an aspect of physical fitness and presented the methodology and percentiles for his testing system. Although Cureton's paper was published over 60 years ago, his comment that flexibility exercises have never been studied intensively because some of the necessary aspects are not measurable in living human subjects still holds true today.

As observed by Gurewitsch and O'Neill (1941) in their study of more than 500 children, Kenny had already assessed muscle spasm in suspected cases of infantile paralysis by asking the children to execute three simple maneuvers that, she supposed, all normal individuals would be able to perform:

1. Straight leg raising to 90°
2. Bending forward with straight knees and touching the floor with the fingertips
3. Bending forward while sitting with the knees straight until the forehead reaches the patella

While her supposition proved to be very optimistic, it provided a fundamental basis for different protocols, such as those of Kraus and Hirschland (1954) and Wells and Dillon (1952), as well as straight leg raising (Hershler and Milner 1980b; Gajdosik, LeVeau, and Bohannon 1985).

From 1948 to 1958, many published papers proposed standardization of range of joint motion measurements in both healthy and sick individuals (Darcus and Salter 1953; Dorinson and Wagner 1948; Dunham 1949; Duvall 1948; Nemethi 1953; Salter 1955; Schenker 1956; Steel and Tomlinson 1958; Storms 1955). In a study published in 1948, Kendall and Kendall managed to have all physical education instructors in the Baltimore, Maryland, educational system collect data on two field flexibility tests providing yes-or-no results. Data from about 5,000 subjects, proportionally distributed from ages 1 to 22 years in both genders, allowed them to present population-curve results for toe touching and bending the forehead toward the knees to reach the patella (Kendall and Kendall 1948). Their data showed that it is very unusual for children older than 4 years of age to be able to touch their forehead to their straight knees and that the smallest percentage of subjects able to touch their toes with their fingertips was among 11- and 12-year-old children.

Some of the most significant contributions in the area of flexibility testing resulted from work derived from Moore's master's thesis. In three articles sequentially published in *Physical Therapy Review* (Hellebrandt, Duvall, and Moore 1949; Moore 1949a, 1949b), Moore, a physiotherapist, presented a broad review of the previous literature, detailed the use of goniometry for range of joint motion measurements, and formally introduced testing and reliability issues. She (1949b) also pointed out very clearly the huge burden caused by the different terminology and reference marks used within the field and, more specifically, the need for standardization in a reference system for range of motion measurements in degrees, for which there are many different systems (e.g., the geometric 180° of arc and the anatomical position of 0° of motion).

The early 1950s brought two major contributions to the area of flexibility testing, one by Wells and Dillon (1952) and the other by Kraus and Hirschland (1954). Both studies, using large samples, aimed to develop flexibility testing standards for school-age individuals. Kraus had coauthored a previous paper in which the toe-touch testing protocol was described (Weber and Kraus 1949), but the enormous repercussions of his 1954 findings—that European children were fitter than American youngsters—made the later article much more famous (Kraus and Hirschland 1954). Thereafter, the testing protocols became widely known. The studies just mentioned are often cited as the primary sources by investigators unfamiliar with the 1941 work of Gurewitsch and O'Neill (Wells and Dillon 1952; Kraus and Hirschland 1954). In the late 1950s, Leighton, who had earlier developed a device to measure the arc of movement in most joints (1942), published a series of articles (Leighton 1955, 1956, 1957a,

1957b) reporting on flexibility testing conducted by applying the Leighton flexometer technique in college students and athletes of different sport modalities.

Contrasting with the simpler techniques employed in previous studies, technology was introduced to flexibility testing in the early 1960s. Johns and Wright (1962) coauthored a classic paper reporting on the relative importance of various tissues in the wrist joint mobility of anesthetized cats. When Karpovich and Karpovich (1959) introduced the potentiometer for measuring angle during a movement, they initiated the use of electronics to measure joint range of motion. Shortly afterward, the American Academy of Orthopaedic Surgeons (1965) published a booklet describing techniques for measuring and normal standards for range of joint motion. People presenting with general hypermobility began to be formally identified by a scoring system (Carter and Wilkinson 1964) also based on Sutro's original work (1947). Other, more sophisticated yet still simple to perform testing and scoring protocols (Beighton and Hóran 1970) followed. In another relevant contribution at this time, Wright's research group further extended the use of a German protocol known as the Schober method to measurements of trunk range of motion in different planes (Macrae and Wright 1969).

In 1970, Nicholas related a low flexibility profile to injury risk during a competitive season in American college football players, bringing considerable attention to the area of flexibility assessment. In terms of flexibility evaluation, the decades of the 1960s and 1970s were characterized by the conducting of a large number of studies that intended to advance the scientific basis of flexibility testing (Finley and Karpovich 1964; Clayson, Mundale, and Kottke 1966; Holland 1968; Harris 1969b; Kettelkamp et al. 1970; Adrian 1973; Beighton, Solomon, and Soskolne 1973; Allander et al. 1974; Coon et al. 1975; Reynolds 1975; Beals 1976; Ellis, Burton, and Wright 1979; Bohannon, Gajdosik, and LeVeau 1985). Also published were discussions of different approaches for reliability and validity assessment and age- and gender-specific norms for existing and sometimes slightly modified test protocols (Leighton 1955;

Kottke and Mundale 1959; Clayson et al. 1962; Ferlic 1962; Macrae and Wright 1969; Ingerval 1970; Rasmussen and Tovborg-Jensen 1970; Beighton, Solomon, and Soskolne 1973; Clarke 1975; Moll and Wright 1971; Haas, Epps, and Adams 1973; Sarrafian, Melamed, and Goshgarian 1977; Wagner 1977; Penning 1978; Boone and Azen 1979; Wolf et al. 1979; Golberg et al. 1980; Bohannon, Gajdosik, and LeVeau 1985; Youdas et al. 1992).

In the late 1970s, sophisticated statistical techniques were incorporated into flexibility testing studies, including intraclass correlation coefficients and the coefficient of variation for reliability determination (Michels 1983; Stratford et al. 1984; Howe, Thompson, and Wright 1985; Merrit, McLean, and Erickson 1986; Gogia et al. 1987; Siegler et al. 1996; Danis and Mielenz 1997; Fredriksen et al. 1997; Rikli and Jones 1997; Armstrong et al. 1998; Jones et al. 1998; Pellecchia and Bohannon 1998; Sabari et al. 1998). With the exception of a simple and practical scoring system published in the prestigious *New England Journal of Medicine* to be used to detect connective tissue-related hypomobility in diabetic children and adolescents (Rosenbloom et al. 1981; Rosenbloom et al. 1982), no new testing method appearing in a major international publication became widely accepted. Most of the new materials or testing protocols were used only in the original work and were not regularly adopted by other research groups (Noyes et al. 1980; Bower 1982; Fitzgerald et al. 1983; Waugh et al. 1983; Murray et al. 1985; Bonci, Hensal, and Tiorg 1986; Silman, Haskard, and Day 1986; Tucci et al. 1986; Krivickas and Feinberg 1996; Fredriksen et al. 1997; Tyler et al. 1999). The major contribution to the area of range of motion measurement during this time was the publication of an issue of *Clinics in Rheumatic Diseases* fully dedicated to the topic (Badley and Wood 1982; Smith 1982; Wright and Hopkins 1982). As pointed out in the foreword (Wright 1982), there is still substantial room for advancement in range of motion measurement and evaluation.

In the last 20 years, most—if not all—textbooks related to exercise and sport sciences (Johnson and Nelson 1979; Phillips and Hornak 1979; Borms and Van Roy 1996; Fahey, Insel,

▶ Major Historical Contributions— 1895-1982

- ▼ Potter (1895)—measured the cubital angle (carrying angle)
- ▼ Albee and Gilliland (1920)—introduced metrotherapy
- ▼ Cureton (1941)—included flexibility assessment in a fitness testing battery
- ▼ Kendall and Kendall (1948)—performed the first large-scale cross-sectional study on flexibility
- ▼ Moore (1949a, 1949b)—introduced reliability issues in goniometric evaluation of flexibility
- ▼ Wells and Dillon (1952)—formally described the sit-and-reach flexibility testing method
- ▼ Kraus and Hirschland (1954)—presented toe-touch test and comparative results for European and American children
- ▼ Karpovich and Karpovich (1959)—introduced electrogoniometry
- ▼ Carter and Wilkinson (1964)—presented a test protocol for assessment of joint hypermobility
- ▼ American Academy of Orthopaedic Surgeons (1965)—published a booklet of norms for joints' ranges of motion
- ▼ Beighton and Hóran (1970)—presented more-developed criteria for quantifying joint hypermobility
- ▼ Rosenbloom et al. (1981)—presented a simple range of motion assessment method for detecting increased risk for future vascular lesions in adolescent diabetics
- ▼ Wright (1982)—contributed to a full issue of *Clinics in Rheumatic Diseases* dedicated to the measurement of joint motion

and Roth 1999; ACSM 2000) and physical medicine and rehabilitation (Cole 1982; Kottke 1982; Norkin, White, and White 1995) have presented material on flexibility testing. The trend has been to establish an association between flexibility and injury incidence and prevention in both sports and health-related quality of life and fitness, thereby strengthening the relevance of this type of measurement (Fleckenstein, Kirby, and

MacLeod 1988; Krivickas and Feinberg 1996; Decoster et al. 1997; Jackson et al. 1998).

Classification Systems

There are a large variety of flexibility testing techniques and protocols, and they are accompanied by distinctive ways to classify them according to the field of interest. In this section, the major classification systems are briefly reviewed and a new classification system is proposed.

Flexibility has static and dynamic components (Corbin and Noble 1980; Reilly 1981). The *static component* relates to the maximal range of motion, while the *dynamic component* relates to the energy expenditure required to perform a given joint movement, measured as *torque*. Because of difficulties inherent in measuring and interpreting the dynamic component of flexibility, and despite its potential relevance, this is carried out only in exceptional research conditions (Magnusson et al. 1996). Nonetheless, recent experimental evidence has shown a good association between the static and dynamic components of flexibility (Gleim and McHugh 1997).

Because there are two basic types of flexibility (Holland 1968; Clarke 1975; Fahey, Insel, and Roth 1999), a logical first step was to classify flexibility testing for static and dynamic flexibility evaluation. Most of the dynamic testing protocols measure joint stiffness curves, that is, the amount of torque required to move the joint in a predetermined arc of movement (Wright and Johns 1960; Johns and Wright 1962). Dynamic flexibility testing is seldom performed in clinical practice or the sport arena; it is considered primarily a research tool.

To classify the static type of flexibility, Harris (1969a) advocated taking a kinesiological approach, in which body actions are considered to be performed by single or multiple joints moving in one or more anatomical planes. According to Harris, there are two types of measurements:

1. Single joint action
2. Composite joint action

Single joint action measurements quantify, by direct measurement of a limb or portion of the body, excursion when only one joint action is involved, whereas *composite joint action* measures the range of motion when more than one joint or type of action within a single joint is involved. It is, of course, possible to perform a single joint action, such as elbow flexion, but the vast majority of daily and sport activities are composite joint actions.

Johnson and Nelson (1979) also proposed classifying static flexibility tests as two types:

1. Relative
2. Absolute

Relative tests are those designated to be relative to the length or width of a specific body part, and *absolute tests* are those that measure only the performance in relation to an absolute goal, such as the linear distance between the performer's body and the floor in a side split. Johnson and Nelson also mentioned that flexibility measurements may be reported as being linear (presented in inches or centimeters as determined by using a tape or yardstick) or rotary (given in degrees of rotation as obtained with the use of a protractor or similar device).

Maud and Cortez-Cooper (1995) introduced different terminology by designating two testing methods:

1. Indirect
2. Direct

This division is related to the type of measurement performed. *Indirect* methods measure the linear distance between two body parts or segments, while *direct* methods are those in which joint angles—that is, angles between body parts or segments—are measured in degrees.

On the other hand, Reilly (1981) used the expression *performance test* to measure situations that involved more than one joint and had a clearly active component, that is, conditions in which the maximal range of motion was obtained by the subject being tested without any external help or support.

Although most static flexibility tests can be performed in either an active or a passive way, depending on their respective protocols and evaluation goals, it should be kept in mind that results and subsequent interpretation vary considerably, especially in terms of limiting factors for maximal range.

Thus far, we have discussed some aspects of the approach to flexibility testing. There are many other relevant aspects still to be considered. For instance, assessment methods vary in terms of the number of joints and movements measured, the instrumentation required, and the length of time needed to complete the test, as well as in statistical aspects. Issues of inter- and intra-evaluator reliability and validity as well as other statistical properties of the results should be known because they are related to the type of scale measurement and the data's inherent distribution characteristics. The presence or absence of "ceiling" and "floor" effects, in addition to non-Gaussian data distribution properties, should be identified; they could be critical to choosing the most appropriate method for a particular testing situation.

After considering the need to include all of these aspects in flexibility assessment, we

> The many different classification systems for static flexibility testing are distinguished from each other according to the type of joint action and the ways in which the measurements are carried out.

developed our own classification of flexibility testing protocols. We integrated into our system some of the concepts presented by other authors and then went further by incorporating clinical flexibility testing in exercise and sports medicine, rheumatology, and related areas into a single classification system. In measurement and evaluation fields, there are additional criteria to be considered in classification or taxonomic systems, including cost and safety issues. However, because costs are somewhat related to instrumentation and the amount of time required to complete (both of which were already incorporated in the proposed model) and because significant health risks are rarely a concern in flexibility assessment, formal inclusion of these items was disregarded. Our system of classification considers 18 major items of evaluation technique or protocol, divided into eight methodological, four operational, and six science-related criteria, as presented

in table 3.1 and briefly discussed in the text that follows.

▶ 1   Type of Flexibility

Range of motion testing could take into account the mechanical forces that resist passive or active movement. This type of flexibility is classified as *dynamic,* evaluating the maximal passive arc of movement, or as *static,* evaluating the maximal range of movement. In reality, the expression *static* must be contextualized, because movement during test performance is not completely static; some movement, normally at a very slow speed, is required to move body segments or parts in order to achieve a maximal range.

▶ 2   Mode of Execution

There are three modes of execution in flexibility testing: active, passive, and combined. *Active execution* is defined by a situation during which no external help is provided; *passive execution* systematically involves either the use of some implement or device or the assistance of an evaluator to achieve the movement. For most, if not all, joint movements, the passive arc of motion is wider than the active one and is very much influenced by both the test subject's muscle factors (such as strength, power, and coordination) and motivation. A simple illustration of the distinction between the active and passive ranges of motion is the person with paraplegia who cannot actively extend her knees but has normal passive knee extension. Exceptionally, a method may mix active and passive forms in some movements (Nicholas 1970) and thereby be classified as a combined test mode.

▶ 3   Number of Movements

Flexibility testing methods vary in terms of the number of movements effectively measured. While many of them are *one-movement* methods, such as sit-and-reach (Wells and Dillon 1952) and straight leg raising (Gajdosik et al. 1985), other protocols incorporate *multiple body movements* (Leighton 1955; Nicholas 1970).

No.	Criterion	Classification
Methodological		
1	Type of flexibility:	Dynamic or static
2	Mode of execution:	Active, passive, or mixed
3	Number of movements:	Single or multiple
4	Number of joints per test item:	Single or composite
5	Number of joint movements in a given item:	Single or composite
6	Total number of joints measured:	Single or multiple
7	Total number of joint movements measured:	Small, regular, and large
8	Global scoring capability:	Yes or no
Operational		
9	Instruments required:	None, simple, or complex
10	Evaluation time required:	Short, regular, or long
11	Feasibility:	Very low, low, regular, high, or very high
12	Measurement unit:	Angular, linear, or dimensionless
Scientific		
13	Reliability:	Low, regular, or high
14	Stability:	Low, regular, or high
15	Validity:	Low, regular, or high
16	Discriminatory power or sensitivity:	Very low, low, regular, high, or very high
17	Applicability:	Very low, low, regular, high, or very high
18	Data distribution characteristics:	Parametrics or nonparametrics

Table 3.1 — Eighteen-Criteria Classification System of Flexibility Testing Methods

4 Number of Joints per Test Item

Joint Flexibility Specificity: Human body movements are performed by *single* or *composite joints* (Harris 1969a). Flexibility testing protocols are designed to measure either single or multiple joints simultaneously.

5 Number of Joint Movements in a Given Test Item

Movement Flexibility Specificity: If a given flexibility testing protocol is able to measure a single movement in a given joint, it can be classified as *single movement,* as in, for example, isolated knee extension. Methods that include *multiple movements* in a single test item are classified as *composite*. An example of multiple movement is the measurement of a full arc of elbow movement in a single plane simultaneously encompassing elbow flexion and knee movements.

6   Total Number of Joints Measured

Most flexibility testing protocols allow range of motion to be measured separately in more than one joint. Therefore, it is possible to label them differently, according to this variable, as *single* or *multiple*.

7   Total Number of Joint Movements Measured

In line with the concept presented in the previous item, flexibility assessment methods vary according to the number of joints being considered: *small* (just one), *regular* (two to five), and *large* (more than five).

8   Global Scoring Capability

Although flexibility is a very complex and specific trait, it is often advantageous to be able to obtain an overall or global profile of an individual's body flexibility by combining the results of range of motion measurements in many different joint movements. Flexibility evaluation protocols can therefore be classified in terms of global scoring possibility as either *yes* or *no*.

9   Instruments Required

Some flexibility testing protocols are equipment free, but others require such sophisticated devices that they are only performed in research conditions. Therefore, it is possible to arbitrarily classify methods according to their equipment requirements into three levels: *none*, *simple* (e.g., a goniometer), or *complex* (e.g., a computerized electrogoniometer).

10   Evaluation Time Required

In clinical practice and population-screening routines, time is a crucial issue, but in advanced research it is usually irrelevant. Flexibility testing methods can be classified into three different categories according to the length of time needed to complete the evaluation: *short* (less than one minute), *regular* (one to five minutes), and *long* (more than five minutes).

11   Feasibility

In schools, health and fitness facilities, and office settings, space availability, material resources, and lighting and climate-control systems may be either very limited or absent, making it difficult to utilize operationally complex evaluation protocols. Moreover, protocols that require special gadgets or vests to be used or the subject to undress could pose additional limitations. In addition, the need to adopt very complicated and physically demanding active and even passive body positions for evaluation has the potential to restrict the usefulness of a given assessment method. Flexibility measurement methodologies may be arbitrarily divided into five levels of feasibility: *very low, low, regular, high,* and *very high*.

12 Measurement Unit

Flexibility tests may be classified in three mutually exclusive categories according to the unit of measurement used: *angular, linear,* and *dimensionless*. All testing protocols whose results are expressed in degrees are classified as angular, while tests providing results in metric units are called linear. The rest, which normally present their results in ordinal or nominal scales of measurement, whether as points or yes-or-no types of responses, are termed dimensionless.

13 Reliability

To be of practical utility, all measurement techniques must be reliable. In terms of intraclass correlation coefficients or kappa statistics (whichever is appropriate for the measurement scale of the protocol), flexibility tests may be classified in one of three levels of reliability: *low* ($r < 0.40$), *regular* ($0.40 < r < 0.75$), or *high* ($r > 0.75$).

14 Stability

In addition to showing high reliability between measurements collected by the same or different evaluators, a good flexibility testing protocol should be stable—that is, the results should have limited variability when all external factors, such as temperature, previous physical activity, and period of the day, are controlled. In reference to the variation between results obtained on two different days, flexibility testing methods can be classified into three levels of stability: *high* ($< 5\%$), *regular* (5 to 10%), and *low* ($> 10\%$).

15 Validity

The capability to truly measure the variable being considered is called validity. While there are different aspects of validity, there is general agreement that all testing methods must be valid if they are to be utilized in practice or research. Since there is not a real gold standard in flexibility testing, it is not feasible to directly determine the validity of a given flexibility testing tool or method. However, it is possible to estimate construct and current validity for most methods, the latter being calculated by product-moment correlation coefficients in comparison with other "reference" protocols. Flexibility testing methods may be somewhat arbitrarily classified in terms of these aspects of validity as *low* ($r \le 0.4$), *regular* ($0.4 < r < 0.6$), or *high* ($r \ge 0.6$).

16 Discriminatory Power or Sensitivity

An assessment technique should be able to detect real differences wherever they exist, in either cross-sectional or interventional approaches. This property supersedes the pertinence of score range, ideally without ceiling or floor measurement effects—extreme scores are very rarely found. The sensitivity of the technique in identifying interventional effects is also influenced by the level of flexibility in the population being assessed. In light of this potential limitation, it seems sufficient to rate protocols for flexibility testing according to their discriminatory power in five arbitrary levels: *very low, low, regular, high,* and *very high.*

17 Applicability

Some aspects of a protocol may interfere with the spectrum of individuals or conditions in which that protocol of flexibility testing can be employed. Ideally, an evaluation method should be valid for all populations, including children, adults, and elderly people of both genders, whether healthy or diseased or at differing levels of regular physical activity that range from sedentary to participation in elite competitive sports. In addition, it should be able to evaluate individuals at the extremes of the flexibility scales, that is, those with hypo- or hypermobility. The applicability of a flexibility test may be considered: *very low, low, regular, high,* or *very high.*

18 Data Distribution Characteristics

Parametric statistics are much more powerful and easier to use and interpret than nonparametric statistical methods are. Nevertheless, some basic assumptions must be met before parametric statistics can be used, namely a high level in the measurement scale (i.e., at least interval-type scoring) and normal or Gaussian data distribution. In this sense, flexibility test routines may be divided by statistical data handling into *parametrics* and *nonparametrics.*

Review of Existing Flexibility Testing Methods

With our classification system in mind, we have the framework for discussing the major flexibility testing methods (chapter 8 will use the 18-point classification system to compare flexibility methods in further detail). If the most important criterion should be chosen for didactic grouping reasons, measurement unit criteria (point 12) is the most appropriate. In this context, the major historical protocols for flexibility evaluation are presented in this section.

Angular Methods

Angular methods for range of motion measurement are those that provide results in degrees of arc—for example, when the angle formed by the longitudinal axis of the forearm and arm is measured during elbow flexion. Goniometry, the measurement of angles, is the term commonly used to identify these methods. Angular methods can be used in either active or passive versions, although they are most often actively applied because at least two trained evaluators are needed for well-performed passive goniometry.

Historically, angular methods were the first to be developed in response to the health-care system's demand for methods for evaluating disability and treatment results in injured World War I veterans (Albee and Gilliland 1920; Gilliland 1921; Glanville and Kreezer 1937; Elward 1939). A large body of knowledge has been accumulated on this topic and books (Norkin, White, and White 1995; Clarkson 1999), book chapters (Maud and Cortez-Cooper 1995; Borms and Van Roy 1996), monographs (AAOS 1965; Clarke 1975; Wright 1982), and review articles (Duvall 1948; Moore 1949b; Storms 1955) extensively cover the methods and techniques available for the angular assessment of joint range of motion. For practically all the body's joints, normal range of motion limits or norms are available (Moore 1949a; Salter 1955; Leighton 1956; Kottke and Mundale 1959; AAOS 1965; Boone and Azen 1979; Bell and Hoshizaki 1981; Fitzgerald et al. 1983; Woods 1985; Pellecchia and Bohannon 1998; Fahey, Insel, and Roth 1999), many of them age and gender specific (Leighton 1955; Clayson et al. 1962; Coon et al. 1975; Bower 1982), allowing the evaluator to comprehensively and appropriately appraise the results obtained. Careful attention to these descriptions is fundamental to ensuring high measurement reliability (Brown and Miller 1998). Possible movements in each of the major joints and corresponding terminology are also available in other source materials (AAOS 1965; Norkin, White, and White 1995; Clarkson 1999). It should also be stated that goniometric measurements are difficult or even impossible to carry out in some joints or specific situations, especially when global movement results in tiny motions of small joints, as occurs in trunk flexion (Buck et al. 1959). For the sake of standardization in this book, some terms for movement description were adopted, including *ankle dorsiflexion* and *plantar flexion* to correspond to ankle flexion and extension, *lateral* and *medial rotation* to correspond to external and internal rotation, and *hyperextension* to describe the exceeding of a joint's neutral position. An important exception to this rule occurs for trunk movements, where *hyperextension* is simply replaced with *extension*.

There are at least three different modes of *arc of movement measurement*, depending on the reference system used (Moore 1949a). In *classical goniometry* (AAOS 1965), the anatomical erect position is considered as 0° of motion. Measurements in each arc logically begin from this value and progress toward 180°, with the only exceptions being the magnitudes of pronation and supination, for each of which the midpoint of the movement is considered the starting position and motion is graded from 0 to 90° in both directions. *Range of motion limitation* is associated with small numerical angular values and both diagnostic and treatment interventions can be easily pursued. While practical, this approach clearly contradicts classical geometry (Moore 1949a), and some professionals have preferred to use a system in which a flexion movement approaches 0° and a fully extended joint reaches 180°. In contrast to classical goniometry, low mobility values are associated with large degree values. A third approach measures degrees for only a *joint arc of movement in one given plane* (Moore 1949a; Leighton 1956), rather than considering the relative contribution of each isolated action. For example, 130° of elbow arc of movement may correspond to 130° of flexion from a neutral extended position or to a combination of flexion and extension limit of amplitude points, such as 10° of elbow hyperextension and 120° degrees of elbow flexion, with each scenario having different clinical significance (Buck et al. 1959). It should be emphasized that angular measurements obtained in different systems are not interchangeable because of the serious risk of data misinterpretation. The system that is chosen should be clearly stated in the data report (Moore 1949a; Buck et al. 1959; Kottke 1983). In addition, none of these systems allow for quantifying small concomitant translations, which quite frequently occur during supposedly single-plane joint movements (Buck et al. 1959).

> Angular methods for flexibility measurement quantify range of motion in degrees by using specially designed instruments.

Having clear instructions for subject positioning, relaxation, and alignment and appropriate lighting conditions are critical in angular measurement (Kottke 1983; Moore 1949b). Good anatomical and kinesiological knowledge is also mandatory to correctly identify the landmarks needed for measuring. It may be wise to use a pen to mark landmarks previously identified, especially the fulcrum, which is the landmark that is considered to be representative of the axis of movement. In population screening or large sample studies, only selected angular measurements are undertaken, normally on the right side of the body (Kottke 1983; Brown and Miller 1998). However, in clinical cases, it is often useful to compare bilateral joint arcs of movement so suspected mobility impairment in a given joint can be compared with a supposedly normal contralateral joint, thereby avoiding bias and other problems associated with the use of reference norms (Moore 1949b). Sometimes,

the same arc of movement measurement performed in different body positions provides distinct values (Sabari et al. 1998), which may be helpful in the clinical differential diagnosis of motion-limiting factors.

Leighton's angular measurement technique (1956) has been widely used by physical educators and athletic trainers, possibly due to the ready availability of reference data from college-age men and some sports modalities (Leighton 1955, 1957a, 1957b) and to the norms recently made available for male and female subjects from 18 to 88 years of age (Brown and Miller 1998). In the original description of the technique (Leighton 1956), the angles of a total of 30 joint arcs of movement performed on both sides of the body and distributed over the upper and lower limbs and trunk were actively measured.

On the other hand, physiotherapists adopted a straight-leg-raising test (Gajdosik and Lusin 1983; Hsieh, Walker, and Gillis 1983; Bohannon, Gajdosik, and LeVeau 1985) to evaluate hamstring tightness in low-back pain patients. Richard Bohannon (Bohannon, Gajdosik, and LeVeau 1985; Gajdosik, LeVeau, and Bohannon 1985), a physiotherapist, has contributed significantly to the understanding of the effects of different joints on straight-leg-raising test results. Because it normally is used in conjunction with other angular measurement methods, the straight-leg-raising test may be performed in either the active or passive version, each of which has a different normalcy value (Bohannon, Gajdosik, and LeVeau 1985), and in one or two legs.

Linear Methods

Linear methods of flexibility testing do not directly measure angles between bony segments, but rather express results in terms of a graduated distance scale, typically in units of centimeters or inches. The first studies on linear methods were published about 20 or 30 years after earlier goniometric descriptive papers and were mainly devoted to flexibility evaluation in children.

In 1941, two independent although somewhat similar protocols for linear assessment of flexibility, both involving basically four distinct movements, were published (Cureton 1941; Gurewitsch and O'Neill 1941). Both studies provided age reference norms for their tests. Gurewitsch and O'Neill's procedures focused on low-back and hamstring flexibility. Cureton's flexibility testing battery evaluated trunk flexion and extension, shoulder flexion, and ankle arc of movement; it made it possible to determine an overall flexibility score by averaging the percentiles of individual movements.

These testing protocols were followed by two other important and well-known contributions to flexibility testing (Weber and Kraus 1949; Wells and Dillon 1952; Kraus and Hirschland 1954). In 1952, Wells and Dillon introduced the *sit-and-reach flexibility test*, aiming to assess hamstring and low-back flexibility. For testing, as the name clearly suggests, the subject adopts a sitting position with the knees fully extended and the feet fixed against an immovable object that has a stick attached and extending beyond the edge. The subject actively bends forward and with arms straight reaches as far as possible toward the stick; then the linear distance between the feet and fingertips is recorded (Wells and Dillon 1952; Corbin and Noble 1980; Reilly 1981). In the original description (Wells and Dillon 1952), subjects had previously warmed up and a total of eight trials were made in order to identify the best score.

There are many versions of the sit-and-reach test, including the modified, back-saver, and YMCA, that differ slightly in methodology and reference norms (Looney and Plowman 1990; Maud and Cortez-Cooper 1995; Schmidt 1995; Cornbleet and Woolsey 1996; AAHPERD 1988; Brown and Miller 1998; Jackson et al. 1998; Cooper Institute for Aerobic Research 1999; Hui and Yuen 2000). Three major changes were made to the original test in these versions. First, the scale of measurement was changed to allow only positive scores by having a 15-inch mark represent the point at which the subject's fingertips are in line with his toes (Cornbleet and Woolsey 1996; Jackson et al. 1998). The second major adaptation was establishing a zero starting point by having the subject reach for the stick while the buttocks, back, and shoulders were kept in contact with a wall; the magnitude of trunk flexion—that is, displacement of the

fingertips in relation to the zero point—was measured (Maud and Cortez-Cooper 1995). The third adaptation was the back-saver sit-and-reach test. The subject is seated on a bench with the knees flexed at right angles and the feet supported on the floor while the trunk is flexed and the arms are fully extended to reach as far as possible, thereby lessening the influence of the hamstrings on performance (Hui and Yuen 2000).

A recent and very interesting study by Holt, Pelham, and Burke (1999) considerably extends the original sit-and-reach test by incorporating a passive form and a total of six different ways (three positions in two modes of execution) to perform it: from the original position, with the knees fully extended and the hips laterally rotated, and as just described but with the hips and knees flexed, the latter at 145°. By using this very creative approach to isolate the contribution of the hamstrings in trunk flexion with the legs fully extended, the authors claimed that they could distinguish the factors limiting movement.

> The main linear methods for flexibility measurement are the toe-touch and sit-and-reach tests, which express their results in inches or centimeters.

The second most popular linear flexibility test is the *toe touch* (Kraus and Hirschland 1954). The subject starts in a standing position with the knees fully extended and then tries to touch the toes with the fingertips. This approach attempts to assess trunk flexion mobility and may measure hamstring extensibility as well. Although in the original description the results were evaluated in a pass-or-fail dichotomy, later studies adopted an elevated bench and incorporated the measurement of the remaining or exceeded linear distance in centimeters or inches in relationship to the toes, which were considered the zero reference mark (Kippers and Parker 1987; Kuo et al. 1997).

Another way of assessing trunk mobility was described in two studies from the same research group and based on Schober's original work (Macrae and Wright 1969; Moll and Wright 1971). After precisely identifying anatomical landmarks in the spine, reference points are marked with the subject in the anatomical position. A simple flexible tape is used for linear measurements, and differences due to lateral or anterior trunk flexion are recorded and evaluated (Moll and Wright 1971).

Some earlier articles and book chapters on the measurement and evaluation of flexibility advocated using other linear testing procedures for other major body joints (Clarke 1975; Johnson and Nelson 1979). However, the vast majority of these active flexibility testing protocols, despite claims of their veracity, have not been validated and attend to very specific needs in sports or dance that today are considered outdated (Cornbleet and Woolsey 1996; Johnson and Nelson 1979).

Dimensionless Methods

Dimensionless methods are those that use neither angular nor linear units of measurement. There are many examples of dimensionless types of measurements in exercise science and medicine; one of these is the widely used international method of appraising a newborn's health by determining its Apgar score (Feinstein 1999). This section will review the most common dimensionless scoring systems for hypo- and hypermobility assessment.

In 1964, Carter and Wilkinson reported data obtained from 285 male and female children ages 6 to 11 years by using a *five-item mobility assessment protocol*. The children's scores were compared with scores obtained in a random sample of 91 boys and girls of similar age who were born with dislocation of the hip (Carter and Wilkinson 1964). Hypermobility, which was diagnosed when three or more of the five items were present, was found to occur about five times more often in children with previous hip dislocation than in typical schoolchildren. By statistically manipulating the original data and calculating the chi-square statistic in contingency tables, we confirmed the existence of a significant difference between the groups at a 1% probability level.

In 1969, Beighton and Hóran published a paper describing hypermobility findings in 100 Ehlers-Danlos syndrome patients studied in two London hospitals. In this study, the authors proposed a modification to Carter and Wilkinson's original scoring system to make measurements in paired joints and increase the maximal scoring from 5 to 9. Articular mobility in a typical African population of over 1,000 subjects was evaluated on the same 0-9 scale, but by using a slightly different scoring system (Beighton, Solomon, and Soskolne 1973). This scoring system has since been used in various papers and quoted in many different ways. Some investigators have designated it simply the *Beighton scale* or *score* (Pountain 1992; Grahame and Pyeritz 1995; Rikken-Bultman, Wellink, and van Dongen 1997), but others have referred to the original contributions describing it by calling it the *Carter-Wilkinson-Beighton scale* (Decoster et al. 1997) or the *Carter and Wilkinson modified by Beighton et al. scale* (Bird, Tribe, and Bacon 1978; Bird, Brodie, and Wright 1979; Biro, Gewanter, and Baum 1983; Seow, Chiow, and Khong 1999).

Of relevance in the use of the dimensionless method is the lack of standardization for a hypermobility cutoff score. Carter-Wilkinson method users agree that the presence of three of the five items characterizes hypermobility (Biro, Gewanter, and Baum 1983; Gedalia et al. 1993), in concordance with the original proposal (Carter and Wilkinson 1964). However, because Beighton, Solomon, and Soskolne (1973) treated mobility scores as a continuum, other authors who have used their method have established a variety of distinct cutoff criteria and terminologies. Grahame and Pyeritz (1995) examined a group of subjects who met strict diagnostic criteria for the Marfan syndrome and determined that most (81%) of the adults had some ($> 1/9$), and 56% had considerable ($> 2/9$) evidence of joint hypermobility.

al-Rawi and al-Rawi (1982) used a 4-point minimum criterion for defining hypermobility in women with genital prolapse, while Bird, Tribe, and Bacon (1978) and Decoster et al. (1997) adhered to a scale that designated a score of >4 out of 9 as diagnostic of hypermobility. Pountain (1992) identified adults as having extreme joint laxity when their scores were between 7 and 9.

The dimensionless method of flexibility assessment was further applied in Nicholas'

(1970) pioneering work with American football players; this method was later adapted in a preparticipation sports evaluation protocol (Goldberg et al. 1980). The Nicholas Flexibility Testing Protocol consists of a visual evaluation of the amplitude of eight body movements, five of which involve the lower part and three of which involve the upper part of the body. Each movement is graded on a scale from 0 to a maximum of 2 in half-point increments according to reference charts presented in the paper (Goldberg et al. 1980). By adding the results, one can obtain values between 0 and 10 for the lower half of the body, between 0 and 6 for the upper part, and between 0 and 16 for overall flexibility, which ranges from maximal tightness at 0 to maximal looseness at 16 (table 3.2). Goldberg et al. found that athletes who scored at the low or high extremes of the scale were more prone to incurring sport injuries during the competitive season, making the protocol useful for defining prevention strategies, which may include stretching or strengthening exercises.

Other dimensionless methods have also been used to identify hypomobility. Pediatric research has demonstrated that children with diabetes mellitus who have stiff and hypomobile hands have an increased risk for developing microvascular disease (Rosenbloom et al. 1981; Rosenbloom et al. 1982). In Rosenbloom et al.'s method (1981), the subjects were asked to tightly approximate the palmar surfaces of the interphalangeal joints of both hands with the fingers fanned. If complete superposition was not achieved, the limitation was confirmed by the examiner, who passively extended the subjects' fingers:

- ▼ Full extension of proximal interphalangeal joints is expected in normal subjects.
- ▼ Distal and proximal metacarpophalangeal joints should extend at least 60°.
- ▼ Wrist and elbow joints should exceed, respectively, 70 and 180° at maximal voluntary extension.
- ▼ Some active range of motion is required in ankle plantar flexion and cervical and thoracolumbar spine lateral flexion.

Joint limitation is classified in four levels:

1. None (includes isolated unilateral or equivocal findings)
2. Mild (bilateral involvement of one or two interphalangeal joints, one large joint, or only the metacarpophalangeal joints)
3. Moderate (bilateral involvement of three or more interphalangeal joints or one finger joint and one large joint)
4. Severe (moderate limitation combined with cervical-spine involvement or obvious hand deformity at rest)

Recently, a new range of motion scale for use in juvenile rheumatoid arthritis has been published (Len et al. 1999). In this dimensionless method for joint motion evaluation, 10 joint movements were graded to four levels ranging from 0 (full movement) to 3 (severe limitation). The scores were inversely associated with the

Table 3.2		Nicholas Flexibility Testing Protocol*	
	Movements	**Normal ranges**
Upper body	Shoulder lateral rotation Shoulder lateral rotation and forearm supination Elbow extension	Boys: 1.5–3.5 Girls: 2.0–4.0
Lower body	Toe touch Knee recurvatum Toe-in Toe-out Lotus position	Boys: 3.0–5.5 Girls: 4.5–7.0

*Each arc of movement is graded from 0 to 2 at 0.5 intervals; all the evaluation reference charts show just three positions corresponding to scores 0, 1, and 2.

ability to handle basic daily living activities (Len et al. 1999). It is quite possible that each of the dimensionless methods discussed here is better suited for assessing specific populations.

> Some flexibility testing methods do not present their values in inches or degrees. We call them dimensionless, and they include the Carter-Wilkinson, Beighton-Hóran, Nicholas, and Rosenbloom testing methods.

Instruments and Devices

Flexibility testing can be carried out in many different ways, from visually judging range of motion to rigorously using sophisticated instruments. This section briefly reviews the most significant contributions to the development of instruments and devices used in flexibility and range of joint motion assessment.

For angular methods of flexibility testing, the *goniometer*—a device that measures angles between bony segments (figure 3.2)—is most commonly used. The origin of the goniometer is literally unknown (Moore 1949a), but was

first mentioned in papers published in the early 1920s (Albee and Gilliland 1920; Gilliand 1921; Conwell 1925). Complete and detailed descriptions and drawings of goniometers used until 1940 can be found elsewhere (Wiechec and Krusen 1939). The goniometer is basically a protractor made of plastic or metal that has at its center two long, slender arms or projections. One or both arms is movable so that it can be placed parallel to the anatomical lever arms of body joints (Moore 1949a).

There are different criteria of classification for goniometers (Moore 1949a; Cole 1982). According to Moore (1949a), there are two basic types of goniometers:

▾ Universal (for measuring angles in different joints)

▾ Specific (designed to measure a given joint motion)

This simple classification was recently updated by Borms and Van Roy (1996) to define the five major types of angle-measuring devices:

▾ Protractor goniometer (Cornbleet and Woolsey 1996)

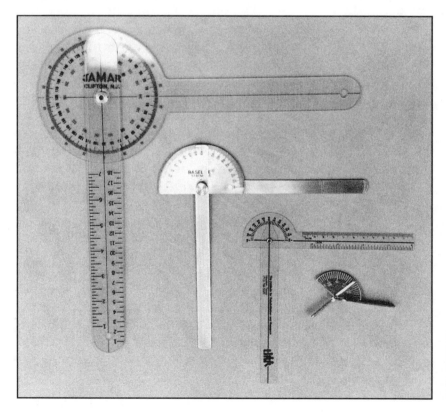

Figure 3.2 Goniometer.
Reprinted, by permission, from P.J. Maud and C. Foster, 1995, *Physiological assessment of human fitness,* (Champaign, IL: Human Kinetics Publishers), 227.

- ▼ Inclinometer (Cornbleet and Woolsey 1996; Saur et al. 1996)

- ▼ Hydrometer (Clayson et al. 1962; Clarkson 1999)

- ▼ Pendulum goniometer (Leighton 1956)

- ▼ Electrogoniometer (Karpovich and Karpovich 1959; Finley and Karpovich 1964; Kettelkamp et al. 1970; Adrian 1973; Hershler and Milner 1980a, 1980b)

Most of the angular measurements in the wrist, elbow, ankle, knee, and hip are performed with protractor goniometers, while the inclinometer and "bubble," or hydrometer, are more often used in trunk and neck evaluations. The pendulum type, exemplified by Leighton's flexometer (Leighton 1956), performs angle measurements in almost all large joints, as well as trunk movements. For the most part, the use of electrogoniometers is restricted to sophisticated clinical or research laboratories (Hershler and Milner 1980a, 1980b).

Quite a few studies have presented modified goniometers or similar devices, most of them

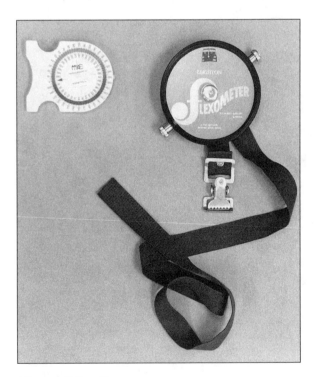

Figure 3.3 Flexometer.
Reprinted, by permission, from P.J. Maud and C. Foster, 1995, *Physiological assessment of human fitness*, (Champaign, IL: Human Kinetics Publishers), 227.

having been tailored for very specific needs (Wainerdi 1952; Schenker 1956; Noer and Pratt 1958; Buck et al. 1959; Weiss 1964; Clayson, Mundale, and Kottke 1966; Wagner 1977; Siegler et al. 1996; Danis and Mielenz 1997), while just one tried to improve protractor readability (Wilmer and Elkins 1947). Considering the complexity and clinical relevance of neck movements, it is not surprising that many studies have proposed techniques and instruments to measure its mobility (Buck et al. 1959; Kadir et al. 1981; Pellecchia and Bohannon 1998). Ellis et al. (1979) undertook an interesting approach that used a very light goniometer that was rigidly affixed to the body of a subject, who was then filmed with a high-speed camera to measure hip function while walking.

A flexometer is a weighted 360° dial and a weighted pointer mounted in a case, both operating freely and independently according to gravitational force (figure 3.3). Independent locking devices are provided for the dial and the pointer, making it possible to stop their movement in any position (Leighton 1956). Commercially available flexometers normally are strapped to the body segment being evaluated to record any movement of more than 20° off the horizontal plane. Despite the fact that the flexometer is portable, low maintenance, simple to use, and reliable when used correctly, its routine use by physical educators in schools and at health and fitness facilities has been remarkably low.

For linear and dimensionless flexibility testing methods, instrumentation is limited or nonexistent. Wells and Dillon (1952) proposed a bench with predetermined dimensions and an attached ruler for the sit-and-reach test, and it is still in use in most settings. For some other linear protocols, only graduated yardsticks or tapes are needed (Cureton 1941; Johnson and Nelson 1979). Dimensionless methods are in general device-independent, sometimes requiring only reference charts or printed evaluation forms. On the other hand, very sophisticated and complex devices are needed for dynamic flexibility evaluation, in which torque and rotational displacement with respect to joint stiffness curves and corresponding hysteresis patterns are recorded and analyzed (Wright and Johns 1960; Wright and Johns 1961).

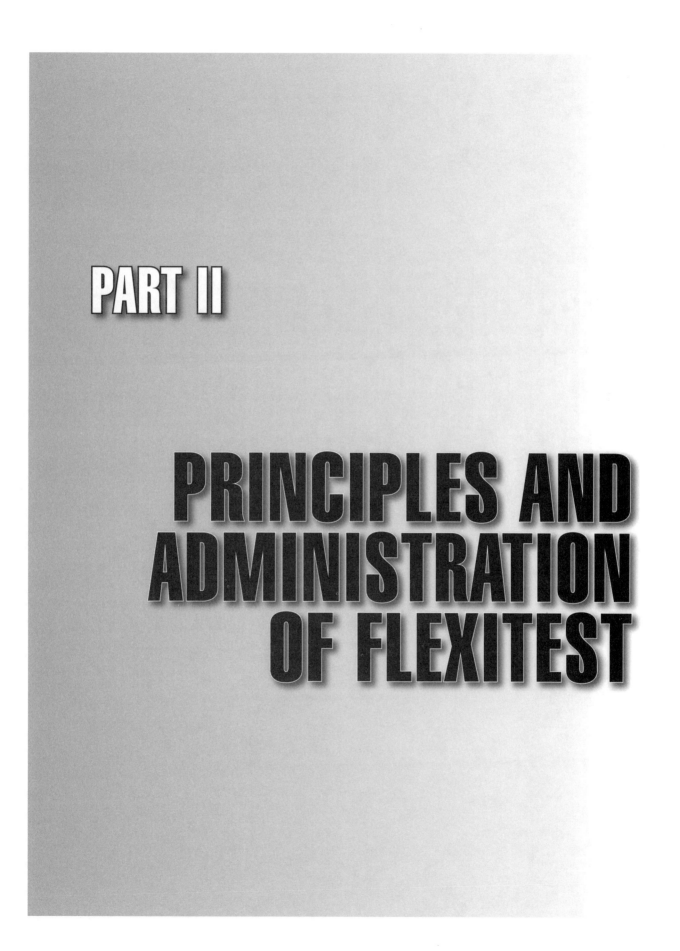

PART II

PRINCIPLES AND ADMINISTRATION OF FLEXITEST

Flexitest Methodology

In part I of this book, an introduction, we presented an overview of flexibility testing and the clinical aspects of physiology, as well as a comprehensive and historical review of the most commonly used linear, angular, and dimensionless tests. In addition, we introduced our own 18-criterion classification system for flexibility testing methods. Part II will cover our testing method called the Flexitest, the main focus of this book. In the present chapter, Flexitest methodology is described and evaluation maps are presented.

General Principles

Flexibility is recognized as an important component of physical fitness, and specific routines for maintaining or improving flexibility levels are included in many of the exercise programs designed for one or more of the different exercise populations. Despite its acknowledged importance, and in contrast to the aerobic and strengthening components of exercise, data discussing the best ways to train or enhance flexibility are limited. Even rarer are sophisticated and scientifically validated routines for assessing flexibility, likely due, in part, to the deficiencies of flexibility assessment methods. It is in this context that the Flexitest fills a specific need in research and practice.

The first version of the Flexitest was proposed by Roberto Pável and myself when we worked together—he as coach and I as physiologist—for a competitive swimming team during the late 1970s. This chapter presents the second and most recent version of the measurement and assessment protocol and includes the evaluation maps, which were redesigned in 1986 and used in my PhD dissertation in 1987 (Araújo 1987).

The Flexitest has been included in the curricula of undergraduate and graduate physical education students since 1980 and has also been taught in a number of physical education professional training courses. Research incorporating the Flexitest has been presented at many local, national, and international congresses and appeared in several languages in a wide variety of international publications in the form of original articles, dissertations, and theses on flexibility (Araújo 1983, 1986, 1987, 1999b, 1999c, 2001; Araújo and Haddad 1985; Araújo and Perez 1985; Farinatti et al. 1995; Farinatti, Soares, and Vanfraechem 1997; Carvalho et al. 1998; Farinatti et al. 1998; Araújo, Pereira, and Farinatti 1998; Coelho and Araújo 2000; Silva, Palma, and Araújo 2000; Chaves, Araújo, and Araújo 2001). The Flexitest has been formally presented in many countries, including the United States, Canada, Belgium, Germany, Slovakia, Puerto Rico, Argentina, Uruguay, Paraguay, Colombia, Ecuador, Costa Rica, and Barbados. Following these scientific publications and presentations, and after being successfully applied in almost 100 Brazilian athletes preparing for the 1988 Olympic Games, the Flexitest began to be routinely used in a large number of fitness centers, gyms, schools, and clubs. In the last decade, the Flexitest has been adopted by exercise and sports medicine doctors for use in their clinics and offices and by the Brazilian military for the flexibility assessment of active personnel.

According to our 18-criterion flexibility test classification system, Flexitest is a *dimensionless test* because results are presented as points, with no linear or angular unit featured. The method involves the measurement and assessment of the maximum passive ranges of motion of 20 joint movements of the body (36 if considered bilaterally), including the main ankle, knee, hip, trunk, wrist, elbow, and shoulder joint movements. Eight movements are made in the lower limbs, three in the trunk, and the remaining nine in the upper limbs. The movements are recorded using Roman numerals in a distal to proximal perspective. Table 4.1 presents the joints tested and simplified kinesiological descriptions of the 20 movements that constitute the Flexitest.

Table 4.1	Kinesiological Description of Flexitest's 20 Movements
Movement	**Kinesiological description**
I	Ankle dorsiflexion
II	Ankle plantar flexion
III	Knee flexion
IV	Knee extension
V	Hip flexion
VI	Hip extension
VII	Hip adduction
VIII	Hip abduction
IX	Trunk flexion
X	Trunk extension
XI	Trunk lateral flexion
XII	Wrist flexion
XIII	Wrist extension
XIV	Elbow flexion
XV	Elbow extension
XVI	Shoulder posterior adduction from a 180° of abduction
XVII	Shoulder posterior adduction or extension
XVIII	Shoulder posterior extension
XIX	*Shoulder lateral rotation at 90° of abduction
XX	*Shoulder medial rotation at 90° of abduction

*With 90° of elbow flexion.

Description of Movements and Evaluation Maps

This section presents in detail the Flexitest and its evaluation-maps methodology, including full descriptions of all 20 movements. For each movement, a kinesiological description and appropriate positions for both subject and evaluator are carefully detailed. Additional comments and helpful hints are also provided, making the evaluation process simple and more reliable. Full compliance with the methodology as presented is extremely important for the proper evaluation and use of the Flexitest data. The written descriptions in this chapter should be used together with and as a complement to the evaluation maps (figures 4.1 to 4.20).

Each movement is progressively graded from 0 to 4 according to the magnitude of the range of motion obtained, as shown in the figure accompanying each movement. The measure is taken passively as the movement is slowly and gradually performed until the maximum range is achieved. This is easily identified by very high mechanical resistance or the discomfort of the subject. The definition of flexibility refers to the maximum physiological range of motion—meaning the maximum range that does not cause lesions—so the comfort of the subject being assessed must be taken into account. Doing so will dramatically reduce the chance of accidents or lesions occurring during the test's application. Once the subject has achieved her maximum amplitude, it is compared with the evaluation map drawn for each of the 20 movements.

A numerical grade is then assigned based on the value shown on the evaluation map that corresponds to the maximal range of motion obtained. For instance, when the amplitude of the subject's movement reaches position 1 on the map, 1 point is awarded until the subject's movement reaches the level corresponding to a score of 2 on the evaluation map, and so on. There are no fractional or intermediate values. It is important to note that an additional point is given only when the subject actually reaches the specified amplitude for the score; even if the range of motion is very close to the next-higher score, it should be rated at the smaller, i.e., already-achieved score.

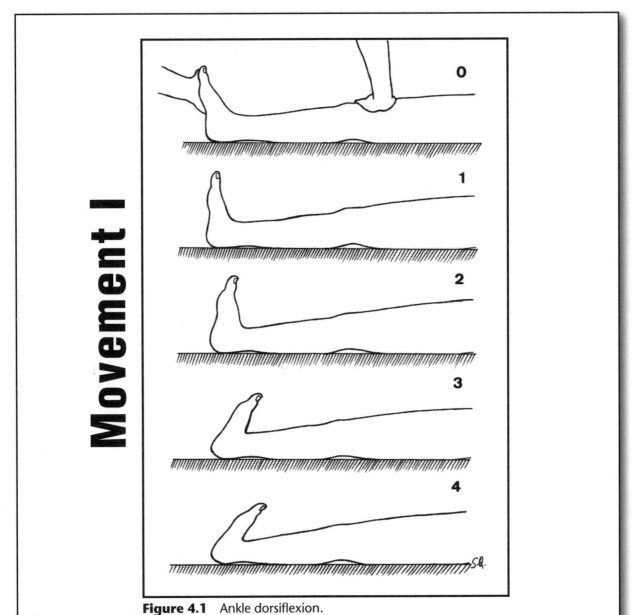

Movement I

Figure 4.1 Ankle dorsiflexion.

▼ **Subject Position**   Supine or seated on the floor with the right leg relaxed and fully extended.

▼ **Evaluator Position**   Kneel perpendicular to the subject. Place your right hand above the subject's right knee. Push the subject's right foot dorsally with your left hand, flexing the ankle by pressing against the metatarsal region while keeping a right angle between your hand and the subject's foot.

*Comments   It is important to eliminate the subject's muscular resistance to motion. Achieving a right angle between the foot and the calf gets a score of 1.

**Hints   It is common for the heel to lift off the floor during the procedure, but this should not affect the evaluation. Have the subject flex the left knee naturally to clear your view of the right leg.

From *Flexitest* by Claudio Gil Soares de Araújo, 2004, Champaign, IL: Human Kinetics.

Movement II

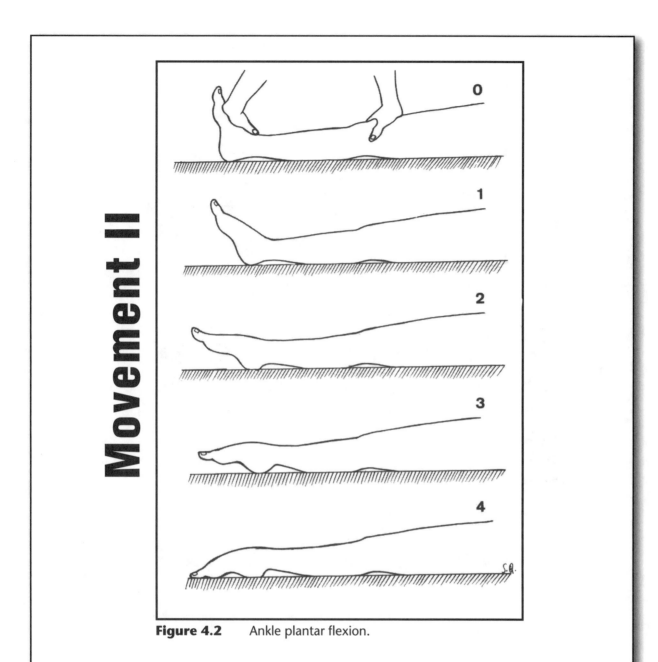

Figure 4.2 Ankle plantar flexion.

▼ **Subject Position** Supine or seated on the floor with the right leg relaxed and fully extended.

▼ **Evaluator Position** Kneel perpendicular to the subject. Place your right hand above the subject's right knee. Place your left hand in the anterior region of the subject's right foot to produce ankle plantar flexion.

***Comments** The position of the subject's toes is not relevant for the measurement. A score of 4 is given when the metatarsal region touches the ground.

****Hints** Attention must be given to keeping the subject's right knee fully extended.

From *Flexitest* by Claudio Gil Soares de Araújo, 2004, Champaign, IL: Human Kinetics.

Figure 4.3 Knee flexion.

▼ **Subject Position**   Prone on the floor with the arms lying naturally above the head and the right knee flexed.

▼ **Evaluator Position**   Kneel beside the subject's left leg and place both hands on the subject's right shin to perform right knee flexion.

*Comments   The posterior thigh and calf do not need to touch to score a 3. To score a 4, it is necessary to gently dislocate the calf laterally in relation to the thigh, which must be performed very slowly and carefully in order to avoid causing ligamentous lesions of the knee structure. (To obtain a score of 4 you are not doing a natural movement, it is almost a dislocation.)

**Hints   Do not consider the subject's right foot position when evaluating the movement. Beware that relatively spastic tight anterior muscles often limit knee flexion, especially in sedentary elderly subjects.

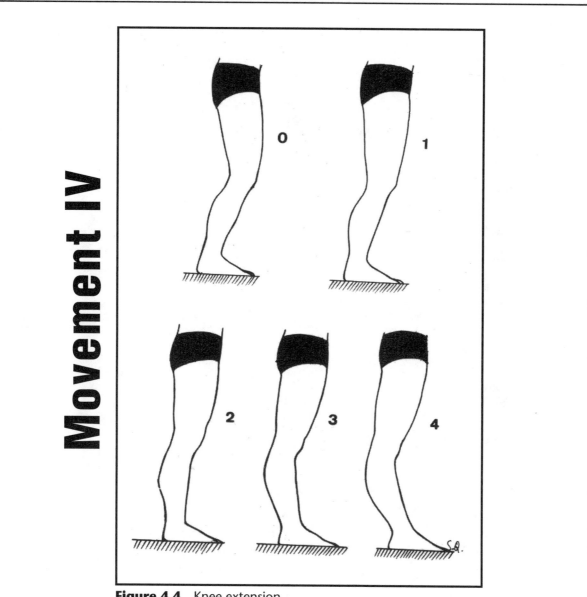

Figure 4.4 Knee extension.

▼ **Subject Position** Standing with the feet together and forcing knee extension without producing hip anteversion.

▼ **Evaluator Position** See comments.

*Comments While this movement is so simple that most subjects can easily perform it without assistance, it is sometimes appropriate for you to help by pushing the thigh immediately above the right knee with your hand. Special attention should be paid to avoiding concomitant hip movements.

**Hints The neutral position corresponds to a score of 2. A score of 4 is clinically denominated *genu recurvatum*.

From *Flexitest* by Claudio Gil Soares de Araújo, 2004, Champaign, IL: Human Kinetics.

Movement V

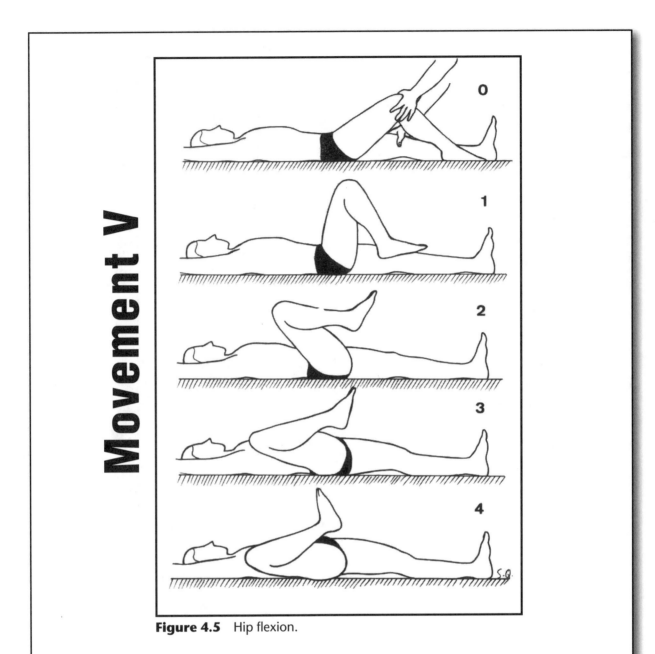

Figure 4.5 Hip flexion.

▼ **Subject Position** Supine on the floor with the arms lying naturally above the head, the left leg fully extended, and the right knee partially flexed.

▼ **Evaluator Position** Stand and keep the subject's left leg extended against the floor by firmly pressing the iliac crest with your right hand while you perform the subject's hip flexion with your left hand on the subject's right shin.

***Comments** In some cases, for convenience, you may use your body weight to help the subject achieve a true maximal passive range of motion (ROM). A score of 3 or 4 can be obtained only if some simultaneous, albeit minimal, hip abduction is allowed.

****Hints** It is very important to avoid hip rotation or contralateral pelvic displacement, which can be easily detected by watching for the left buttock moving away from the floor or by the impossibility of keeping the left iliac crest still.

Movement VI

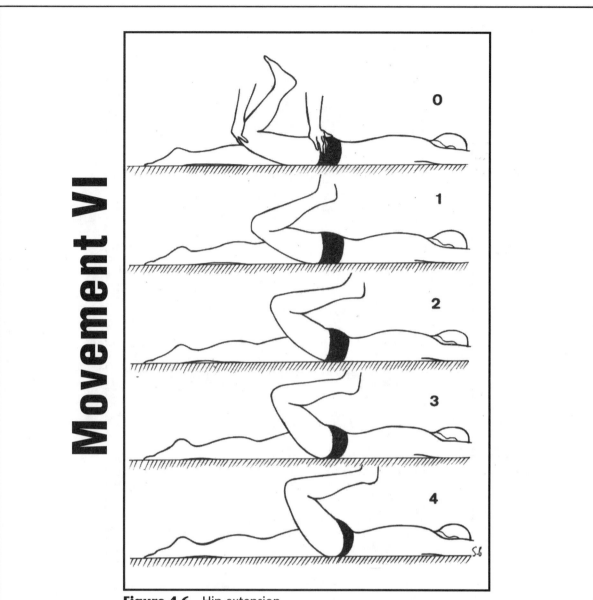

Figure 4.6 Hip extension.

▼ **Subject Position**   Same as in movement III.

▼ **Evaluator Position**   Kneel lateral to the subject and perform right hip extension by placing your left hand under the subject's right knee while pushing the subject's right hip against the floor, impeding the movement, with the palm of your right hand.

*Comments   The major problem in performing this movement is preventing the subject from elevating the right iliac. Again, do not consider the foot's position when evaluating the hip's ROM.

**Hints   Asking the subject to start the movement will facilitate your work.

From *Flexitest* by Claudio Gil Soares de Araújo, 2004, Champaign, IL: Human Kinetics.

Movement VII

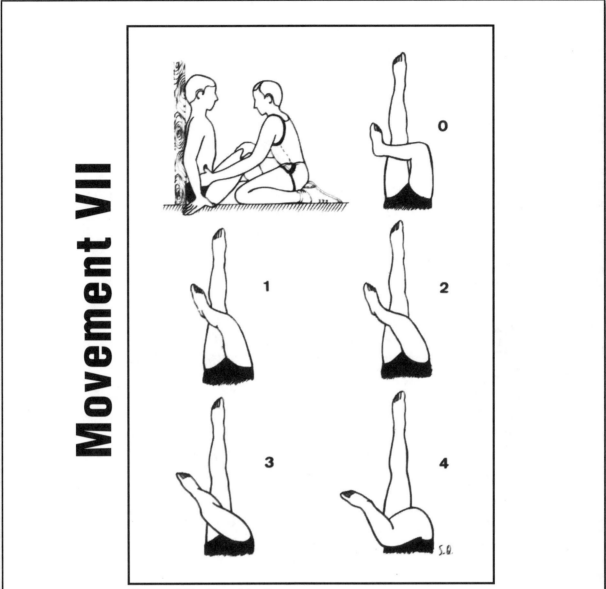

Figure 4.7 Hip adduction.

▼ **Subject Position** Seated on the floor with the trunk and lumbar region remaining as straight as possible, the left leg fully extended, the right knee flexed at about 90 degrees, and executing hip adduction.

▼ **Evaluator Position** Kneel in front of the subject and use your left hand to hold the subject's right hip so that it does not rotate as you perform hip adduction on the subject by placing your right hand on the lateral and distal part of the subject's right thigh.

*Comments It is of utmost importance to prevent the subject's right hip from rotating. The subject's right foot will naturally follow the leg movement, but this movement is not relevant for ROM evaluation. When the subject's right knee reaches his body median line, a score of 2 is awarded, while for a score of 4 there must be complete contact between the medial side of the subject's thigh and chest.

**Hints You may want to have the subject keep his back in contact with a wall or use your left leg as a support. Alternatively, you can ask the subject to put his hands beside his hips to support the trunk and help keep the spinal column straight.

From *Flexitest* by Claudio Gil Soares de Araújo, 2004, Champaign, IL: Human Kinetics.

Movement VIII

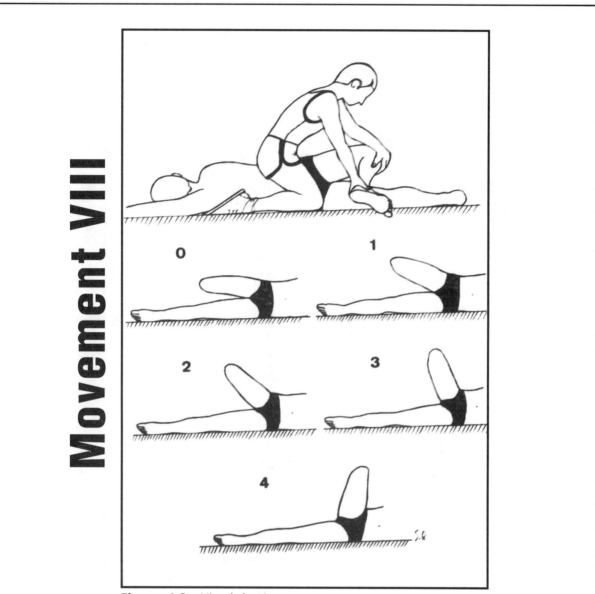

Figure 4.8 Hip abduction.

▼ **Subject Position**   Lying in a left lateral position with the arms extended above the head. The left leg is fully extended and the right leg, with the knee bent and the foot in a natural position, is aligned with the body axis.

▼ **Evaluator Position**   Kneel lateral to the subject to perform the hip abduction. Press your right hand against the subject's right iliac crest to prevent hip rotation while bringing the subject's right leg toward the trunk in a frontal plane with your left hand.

*Comments   Achieving a right angle between the trunk and right thigh corresponds to a score of 3. Attention should be paid to avoiding even minimal hip rotation, which could significantly increase ROM.

**Hints   In order to minimize right hip rotation, insist that the subject keep her left leg fully extended.

From *Flexitest* by Claudio Gil Soares de Araújo, 2004, Champaign, IL: Human Kinetics.

Movement IX

Figure 4.9 Trunk flexion.

▼ **Subject Position** Seated with the legs fully extended and forming a right angle with the trunk. The arms are flexed and the hands are joined at the back of the neck.

▼ **Evaluator Position** Kneel behind the subject and place the palms of both hands under the subject's shoulders with your arms in the supine position.

*Comments It is mandatory that the subject's buttocks remain in contact with the floor and the knees be fully extended during measurement. When the movement is performed in the seated position, as we now recommend, stay behind the subject and push his trunk toward his legs. If the subject cannot reach the starting position without flexing his knees, the measurement is scored as a 0. When only cervical movement is observed, the score is 1, but if there is lumbar motion, the score should be at least a 3. A score of 4 is awarded when the trunk and anterior thigh are completely superposed.

**Hints Ask the subject to initiate the trunk flexion movement to substantially reduce your effort. Do not be distracted by head or cervical mobility; the evaluation should primarily consider the thoracic and lumbar regions of the spine.

From *Flexitest* by Claudio Gil Soares de Araújo, 2004, Champaign, IL: Human Kinetics.

Movement X

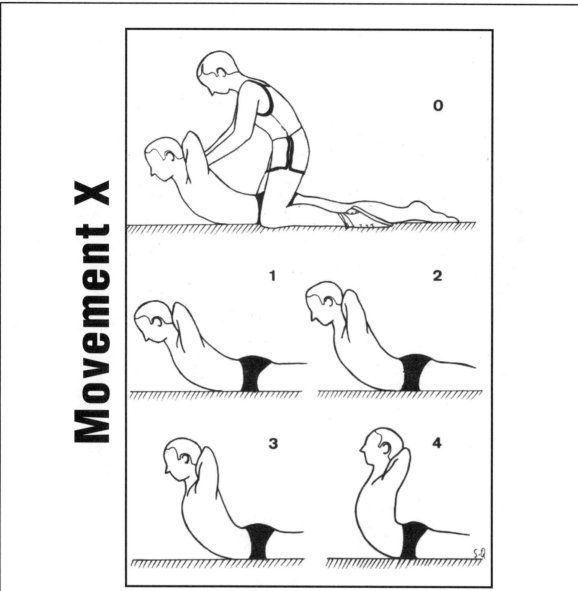

Figure 4.10 Trunk extension.

▼ **Subject Position** Prone with both legs extended and with the hands joined at the back of the neck.

▼ **Evaluator Position** Kneel or stand with your trunk partially flexed, keeping the subject's body between your knees or feet. Execute the subject's trunk extension with your hands placed over the subject's shoulders.

*Comments As suggested in movement IX, ask the subject to actively initiate the movement. For evaluation, reference the trunk extension to avoid the potential confounding and distracting effects of head and arm positions.

**Hints Letting your feet touch the subject's lateral hip area will allow you to more easily detect the iliac crest if it moves away from the floor; or, place a mirror on a lateral wall to monitor the movement.

From *Flexitest* by Claudio Gil Soares de Araújo, 2004, Champaign, IL: Human Kinetics.

Figure 4.11 Trunk lateral flexion.

▼ **Subject Position** Same as in movement X.

▼ **Evaluator Position** Same as in movement X, but place your right hand on the subject's right arm to make lateral trunk flexion easier to perform.

*Comments The subject should perform the movement with no spinal extension, i.e., his chest should be minimally separated from the floor.

**Hints As with the two trunk movements mentioned previously, have the subject start the movement. Also note the spinal curvature when the subject's back is bare for a better assessment.

From *Flexitest* by Claudio Gil Soares de Araújo, 2004, Champaign, IL: Human Kinetics.

Movement XII

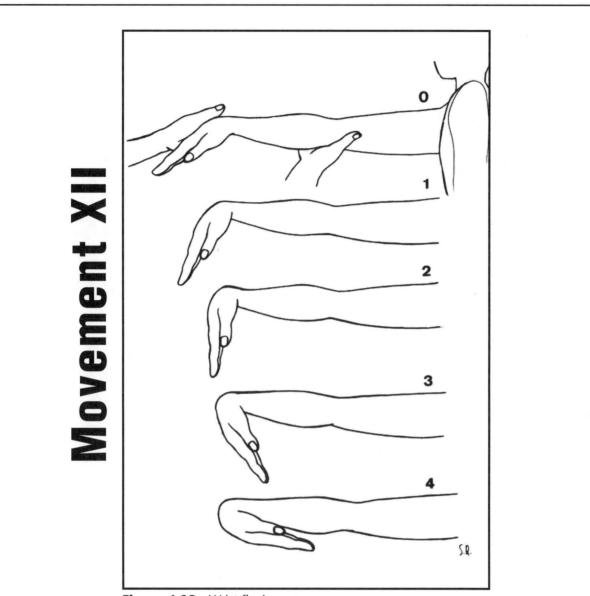

Figure 4.12 Wrist flexion.

▼ **Subject Position**   Standing with the right arm and elbow extending forward in the prone position from the front of the body (at a right angle with the body's major longitudinal axis).

▼ **Evaluator Position**   Stand at the subject's side (medial view), and with your right hand in the supine position keeping the subject's right arm fully extended, perform wrist flexion with the left hand; support the subject's right hand by placing your hand on the subject's posterior metacarpal region to form a right angle between your and your subject's hands.

***Comments**   Do not allow the elbow to flex to make a reliable assessment. The subject's arm should be extended to the front of the body, with no abduction of the corresponding shoulder. Observe the movement from the medial (previously called the *internal*) side of the subject's arm.

****Hints**   The pressure you apply to perform the wrist flexion should not be exerted on the subject's fingers, but rather on the metacarpal region. The positions of the fingers should not be taken into consideration for evaluation purposes.

From *Flexitest* by Claudio Gil Soares de Araújo, 2004, Champaign, IL: Human Kinetics.

Movement XIII

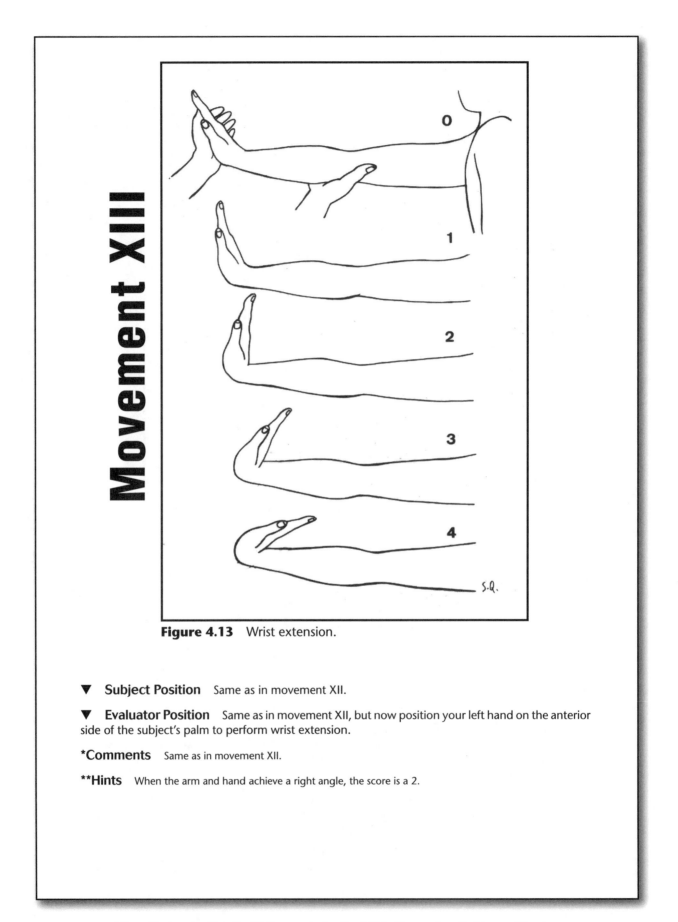

Figure 4.13 Wrist extension.

▼ **Subject Position** Same as in movement XII.

▼ **Evaluator Position** Same as in movement XII, but now position your left hand on the anterior side of the subject's palm to perform wrist extension.

*Comments Same as in movement XII.

**Hints When the arm and hand achieve a right angle, the score is a 2.

From *Flexitest* by Claudio Gil Soares de Araújo, 2004, Champaign, IL: Human Kinetics.

Figure 4.14 Elbow flexion.

▼ **Subject Position** Same as in movements XII and XIII, except that the right elbow is now flexed.

▼ **Evaluator Position** Same as in movements XII and XIII, but now stand on the subject's lateral (previously called *external*) side for a lateral view. Your right hand still will be underneath the elbow, but place your left hand on the distal portion of the subject's forearm to perform the right elbow flexion.

*Comments Complete superposition of the forearm over the arm is rated 3. Observe the movement from the lateral side of the subject's arm.

**Hints For a score of 4, as with movement III (knee flexion), it is necessary to gently displace the forearm laterally in relation to the arm.

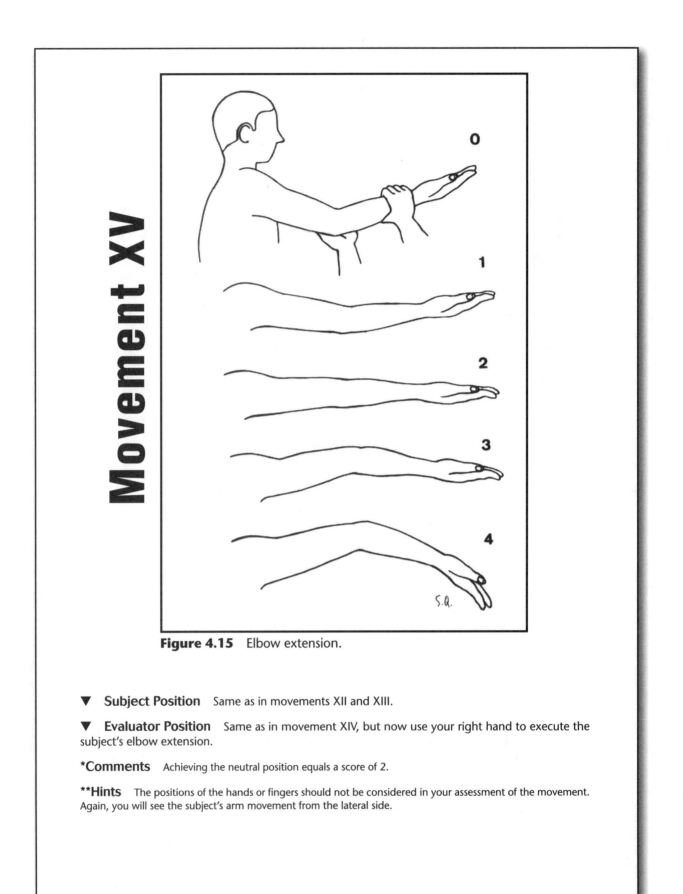

Movement XV

Figure 4.15 Elbow extension.

▼ **Subject Position**   Same as in movements XII and XIII.

▼ **Evaluator Position**   Same as in movement XIV, but now use your right hand to execute the subject's elbow extension.

*****Comments**   Achieving the neutral position equals a score of 2.

******Hints**   The positions of the hands or fingers should not be considered in your assessment of the movement. Again, you will see the subject's arm movement from the lateral side.

Movement XVI

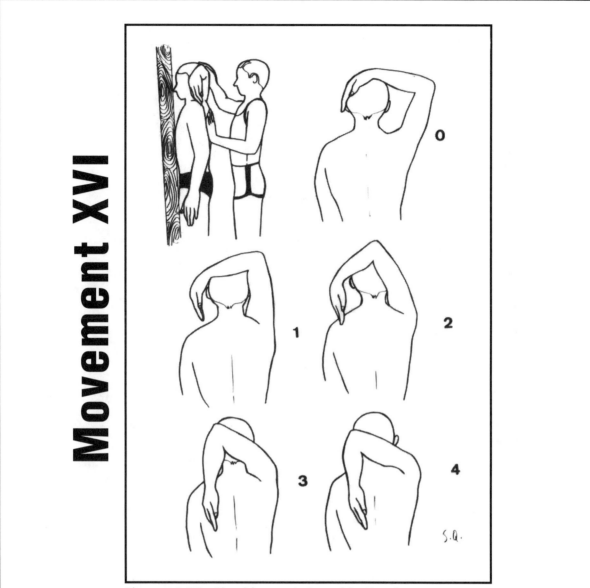

Figure 4.16 Shoulder posterior adduction from a 180° abduction.

▼ **Subject Position** Standing with the head flexed slightly anteriorly and the shoulder in posterior abduction beginning at 180°.

▼ **Evaluator Position** Stand behind the subject and gently push the subject's upper back with your left hand to stabilize it while your right hand, placed on the arm's distal portion, executes the movement.

***Comments** When the subject's right arm is parallel to the body's longitudinal axis, the score is 1. When the right elbow is exactly over the body's median line, the score is 2. The subject should inform you when the maximum ROM is reached. No trunk lateral flexion should occur.

****Hints** You may want to have the subject face and press her chest against a wall. This alternative was included in the original Flexitest description, but now it is only occasionally used.

From *Flexitest* by Claudio Gil Soares de Araújo, 2004, Champaign, IL: Human Kinetics.

Movement XVII

Figure 4.17 Shoulder posterior adduction or extension.

▼ **Subject Position** Prone with the chin on the floor, the legs extended, and the arms abducted and extended, palms pointing toward the floor.

▼ **Evaluator Position** Same as in movements X and XI, but hold the subject's palms in your hands to execute the movement.

*Comments When a right angle is formed by the subject's trunk and arms, the score is a 2. In a subject with normal limb-trunk proportions, when the wrists are superposed, the score is a 3 and when the elbows are superposed, the score is a 4.

**Hints Before starting the movement, ask the subject to relax his arms. Remind the subject to squeeze your hands when the maximum tolerable ROM is reached.

From *Flexitest* by Claudio Gil Soares de Araújo, 2004, Champaign, IL: Human Kinetics.

Movement XVIII

Figure 4.18 Shoulder posterior extension.

▼ **Subject Position**   Same as in movement XVII, although the arms are not abducted.

▼ **Evaluator Position**   Same as in movement XVII. Gently hold the subject's hands to execute the movement.

***Comments**   To begin the movement, you must assume a "zero" position, ensuring that the subject's arms are not abducted. This movement should be performed very slowly to reduce the risk of injury.

****Hints**   Again, have the subject squeeze your hands when the maximum tolerable range of motion is reached. The subject may feel somewhat insecure with this movement, so it is vital that it be performed slowly.

From *Flexitest* by Claudio Gil Soares de Araújo, 2004, Champaign, IL: Human Kinetics.

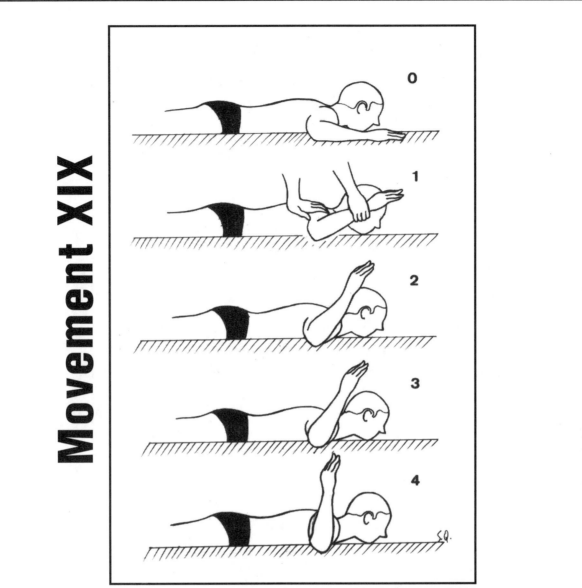

Movement XIX

Figure 4.19 Shoulder lateral rotation with abduction of 90° and elbow flexion of 90°.

▼ **Subject Position**   In the prone position, keeping both shoulders in contact with the floor, with the right arm abducted and the elbow flexed (both at 90 degrees) while the shoulder is on a 90° lateral rotation. The left arm should be placed naturally alongside the body.

▼ **Evaluator Position**   Kneel at the subject's side to execute the movement with your right hand, holding the subject's right forearm near the wrist while placing your left hand between the right acromion and neck to keep the subject's right shoulder against the floor.

*Comments   A very important thing to consider in this assessment is the angle between the subject's right forearm and the body's longitudinal axis, without taking into account the positions of the hand or fingers. Be certain that the subject's right shoulder remains in contact with the floor.

**Hints   Hold the subject's arm firmly, but avoid restricting her shoulder rotation.

From *Flexitest* by Claudio Gil Soares de Araújo, 2004, Champaign, IL: Human Kinetics.

Movement XX

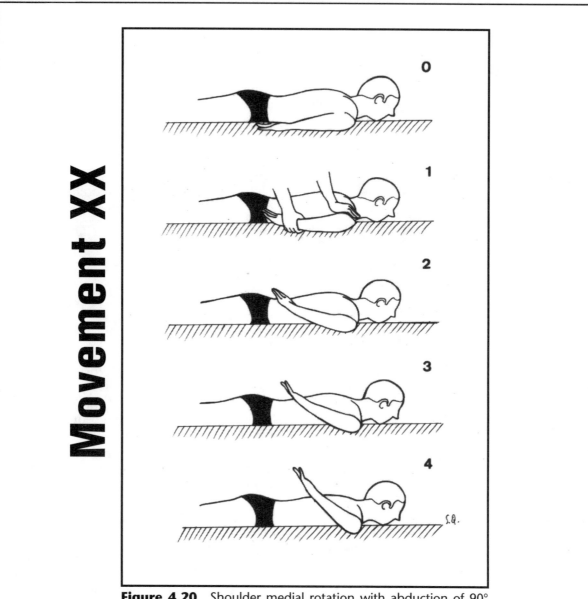

0

1

2

3

4

Figure 4.20 Shoulder medial rotation with abduction of 90° and elbow flexion of 90°.

▼ **Subject Position**   Same as in movement XIX, but place the shoulder in a 90° medial rotation position.

▼ **Evaluator Position**   Same as in movement XIX, but use your right hand to perform medial rotation of the subject's right shoulder.

*Comments   Basically the same as for movement XIX. Not being able to correctly perform the initial position due to limited shoulder mobility represents a score of 0. If you can put your fingers between the floor and the subject's forearm without the subject's elbow leaving the ground, a score of 1 has been achieved.

**Hints   Avoid being distracted in your evaluation by wrist or finger movements.

From *Flexitest* by Claudio Gil Soares de Araújo, 2004, Champaign, IL: Human Kinetics.

▶ Flexitest

Methodological issues

▼ Aim is to assess maximal passive range of motion by visual comparison with evaluation maps.

▼ Test measures ROM in 20 body movements—8 in the lower limb, 3 in the trunk, and 9 in the upper limb.

▼ Typically, administer only on the right side of the body for bilateral movements.

▼ Do not allow warm-up or intense physical activity to take place before measurement.

▼ Each movement is evaluated on a scale of five possible mobility scores (ranging from 0 to 4), with higher scores reflecting larger ranges of motion.

▼ Award the next-higher score only when the subject achieves the range of motion that corresponds with that score as presented on the map.

▼ Measuring takes three to five minutes when a specific sequence and five body positions are used.

There are several considerations to keep in mind in standardizing the application of the Flexitest. Traditionally, the test is applied only on the body's right side; bilateral differences are very rare, but may occur in some specific movements in cases of over- or underuse. Since body temperature influences flexibility, control this variable by avoiding any warm-up or intense physical activity in the hour before the measurements are made. Flexitest application time varies according to the evaluator's experience and the condition of the subject, but typically takes three to five minutes when a modified sequence of movements that minimizes changes of body posture is employed. Later, in chapter 7, we provide examples of scientific studies that have used the Flexitest; this presentation of research and methodological issues will allow you to further understand the test's evolution and scientific merit.

The current Flexitest does not follow the numerical order of movements, sequencing them instead to minimize body-position changes. It involves five positions:

1. Lying in the supine position: movements I, II, and V

2. Lying in the prone position: movements III, VI, X, XI, XVII, XVIII, XIX, and XX

3. Lying in the lateral position: movement VIII

4. Seated: movements VII and IX

5. Standing: movements IV, XII, XIII, XIV, XV, and XVI

In a simplistic analysis, mobility scores for young adults could be evaluated according to the following scale:

0 = very poor

1 = poor

2 = average

3 = good

4 = very good

According to the scale of measurement and the way the evaluation maps were designed, the data tend to follow a Gaussian distribution in young adults, so the central tendency (median and mode) is a score of 2; 1 and 3 are scored less frequently, and the extreme scores, 0 and 4, are quite rare. Therefore, even though Flexitest assessment may and should be done for each of the movements and joints, it is valid and appropriate to add the results of each of the 20 movements to get an overall flexibility or joint mobility index called a *Flexindex*, which is a major advantage of the Flexitest when compared to goniometry. Moreover, with the Gaussian nature of each movement's scale and the global scale, it is possible to study the entire mobility range, because the maximum extreme values—0 and 80 points—although theoretically possible, are never actually obtained. Therefore, there are no "ceiling" or "floor" effects, which make the clinical use and interpretation of simple tests more difficult by reducing their specificity or sensitivity.

Flexitest Administration

In this section we present hints, solutions, and procedures, gathered during our more than 20 years of applying the Flexitest, to facilitate its

▶ Flexindex

- ▼ A meaningful index of global flexibility
- ▼ Equals the sum of all 20 individual mobility scores
- ▼ Scores range from 0 to 80 points
- ▼ Permits intra- (longitudinal) and intersubject (cross-sectional) comparison

use and the practical interpretation of results. This is our "Flexitest recipe" for students and practitioners, divided according to the different aspects of test administration.

Preparing yourself and the place:

1. Acquaint yourself with the evaluation maps. Over time, try to memorize the lower and upper limits for a score of 2 for each movement. Doing so will lessen your evaluation mistakes.

2. Do not overestimate your knowledge, and practice with the evaluation maps. Periodically review the execution positions, observation angles, and scoring for all the movements. This should be done at least once a month after you have quite a bit of experience with the Flexitest, but even more frequently when you have less experience with it.

3. Ask a colleague who is at least as proficient as you are to periodically monitor your administration of a Flexitest evaluation. Unrecognized slips in procedure are easily acquired, and difficult to correct once learned.

4. Establish the physical positions that you and the person being evaluated will take, and maintain them according to the needs of the test.

5. Note if the walls of the room can be used. There might be baseboards, electrical cords, or other objects that make some positions difficult. (Sometimes the back of a door is sufficient for Flexitest administration.)

6. Determine whether there is adequate space for performing the movements. Because the movements require very little space,

this usually is not a problem, but if other people are watching or helping, take their needs into consideration.

7. Note that mirrored rooms are seldom of help in and may even complicate or cloud the assessment of the range of motion for a movement.

8. Verify that the lighting and climatic conditions in the room are appropriate for obtaining and grading the measurements.

Preparing the subject for evaluation:

1. Verify that the subject has not exercised within the hour preceding Flexitest administration.

2. Make certain the subject is comfortable in his clothing and by explaining the procedure in appropriate terms.

3. If the subject wears eyeglasses, have her take them off while performing the Flexitest movements, because they may fall off and even break.

4. Have the subject also remove all jewelry, particularly bracelets, which may restrain wrist extension, and expensive watches, which can be scratched or hit on the floor when the subject executes movement XX.

5. Inform the subject of the positions he will have to assume for the measurements to be taken.

6. Explain the measurement criteria and how the movements, the scores for which range from 0 to 4, are graded (i.e., by comparing the range achieved with the evaluation map).

7. Explain that some degree of discomfort but no injury is to be expected when a movement nears the maximum range of motion, especially in the shoulder and wrist movements.

8. Clearly explain to those who are not particularly flexible that a score of 4 is not what a subject is expected to achieve, and that it may even be a health hazard to be extremely flexible.

9. When measuring children, show them the evaluation maps beforehand or allow them

to watch an evaluation being performed before undergoing their own.

10. Before beginning the test, establish with the subject the ways in which you will communicate so that you will recognize the subject's degree of discomfort or if the maximum range of motion has been reached. One such way is through the use of signs (for instance, ask her to squeeze your hands in movements XVII and XVIII), words, or expressions. Also, pay attention to the subject's facial expressions; they may serve as hints for when to stop the movement and the measurement or, at the least, reveal how she is feeling.

Movement execution aspects:

1. Be conscious of your own body posture when applying the test. Doing this will help prevent evaluation errors and serve as an example for the subject. Poor posture may cause discomfort in the lower back, especially when you must assess a number of subjects in a short period of time or if the subject is heavy.

2. If the subject is tense, postpone the measurement procedure. Talk to him to help him relax as much as possible (this is especially critical for shoulder evaluation, since some of the movements could have a substantially lower range of motion when the related muscles are not completely relaxed).

3. If a wall is not available to support the subject in some of the movements, be careful to ensure that lateral movements that would have been prevented or restrained by the wall do not occur (such as hip and trunk rotation in movement VII, hip adduction).

4. If, when attempting to correctly perform a movement, the minimal range sufficient to score a 2 is barely reached, do not try to go further. It is highly unlikely, and perhaps impossible, that the ROM could reach a score of 3 or 4, and the risk for injury is increased.

5. If a score of 4 is achieved, do not go further and force the amplitude of the movement to see its maximum point. Continuing the movement may cause injury or unnecessary discomfort, and may also compromise the willingness or demeanor of the subject, potentially jeopardizing the performance and evaluation of the remaining movements.

6. For a score of 4 to be given in some movements (e.g., movements III, XIV, and XVIII), a position that is contrary to nature must be assumed; carefully and gradually perform these movements.

7. On the other hand, do not hesitate to stretch the movement by applying force gradually and continuously until the maximum amplitude is reached. With this approach, the potential for the Flexitest to cause injury is extremely low to nonexistent.

8. Do not demand that the subject achieve perfect and complete knee extension in movement IX, trunk flexion. Many subjects cannot perform a full extension, even when only the knees are considered.

9. Follow the map illustrations to position the ankles for some of the lower-limb movements and trunk flexion; extending the ankle to an extreme dorsal or plantar position may change the ROM of the movement being made.

10. Avoid repeating a movement two or more times consecutively when attempting to reach the maximum range of motion, because the results will tend to unpredictably improve. If a movement must be repeated to resolve a question or doubt, do it only after completing all the other movements. Allow a minimum 30-second interval between consecutive measurements of the same movement.

11. If you are to reassess a subject, do not study the previous assessment, because it may affect your judgment and the reliability of your measurements, thus compromising the interpretation of results.

12. Do not be influenced in your evaluation by a subject's having a background of sports participation or intense flexibility training. Often, athletes are less flexible than

untrained individuals of the same age and gender!

13. If you suspect a limitation may be due to injury or physical impairment, make the measurement on the contralateral limb or to the other side (as in movement XI, trunk lateral flexion).

14. Use special caution when helping subjects with a history of sciatic pain, spondylolisthesis, or intervertebral disk disease perform movements X and XI.

15. When measuring movement IX, trunk flexion, ask the subject to place her legs against the wall only if you feel a score of 4 may be reached. Otherwise, make the movement easier to perform by having the subject seated, and closely monitor that the knees are extended and the legs are straight against the floor.

16. Movement XVI can be done by the subject by using his left arm to pull the right. The wall usually will not be needed for support, but be careful not to let the subject make a lateral trunk flexion. Not using the wall may be beneficial, though, because it permits the cervical spine to flex slightly, making the movement easier to perform.

17. Bedridden subjects initially can be assessed in a medical stretcher, changing the proposed sequence of movements, adjusting according to the individual circumstances. If necessary, perform movements XVII and XVIII using the right arm only, with the anterior portion of the left shoulder kept against the stretcher surface.

18. In subjects with very low exercise tolerance (i.e., a functional capacity lower than 4 to 5 METs), the positioning for some of the measurements—in particular, the prone position—may cause shortness of breath, cyanosis, and chest discomfort. If that is the case, movements involving trunk flexion and lateral extension should be done from a standing position with the hips against the lateral face of stretcher while you stand directly behind the subject. Most of these subjects will score a 0 or 1 in these movements and it will be easy to identify mobility limitation.

19. Take your time when performing a movement, and do it uniformly and without interruption. You should be able to easily and comfortably administer the Flexitest in less than five minutes.

20. You may want to ask the subject to actively start the movement and only then take over to reach the highest possible amplitude. This is critical if you weigh much less than your subject.

Recording information:

1. When you say a score aloud to a scorekeeper, keep your tone of voice as neutral as possible and never use positive or negative expressions.

2. If you are working with a scorekeeper, he should know the proper sequence and numbers of the movements and should repeat, in a neutral tone of voice, the score as he enters it.

3. When you finish performing the 20 movements, double-check the scores to be sure that no mistakes occurred, because it will be very difficult to identify and correct them later.

4. If possible, record the results in an electronic spreadsheet that includes the formulas needed to calculate the Flexindex and variability indexes.

5. If there is no scorekeeper, record only the scores that are not 1 for elders, 2 for adults, and 3 for children. This will significantly reduce the time needed to perform the procedure, because it is very likely that only a few scores will have to be written down.

6. Mental calculation of the Flexindex is easy if you consider only the movements with scores other than 2 and start at 40 points. Add 1 point for each score of 3 and 2 points for each score of 4 and, conversely, subtract 1 and 2 points for scores of 1 and 0, respectively. For instance, if the subject scored a 4 on one, 3 on two movements, a 1 on two others, and a 2 on all the rest, we have $40 + (2 \times 1) + (1 \times 2) - (2 \times 1) = 42$ points. For children, record the scores for only the movements that did not rate

a 3 and start at 60 points; then add and subtract as previously described.

Further Comments

Although we recommend that these standardized procedures be adopted, in some cases this is not feasible. There are two types of situations that often demand changes in the Flexitest methodology. The first is when the subject is unable to perform, even passively, one or more of the Flexitest movements due to, e.g., a plaster cast or recent surgery. Very elderly subjects and those with physical disabilities also sometimes have difficulty adopting or tolerating some of the recommended positions. In these cases, you need a special approach that uses different subject positioning or other solutions to assess flexibility. When the problem is restricted to one limb or one side of the body, the measurement can be made on the other side to determine the Flexindex (note in the patient's record which of the measurements were taken on the left side). In the case of subjects having a higher degree of impairment, we have adapted the Flexitest methodology to use three positions:

1. Lying on a bed or stretcher: movements I, II, IV, V, and XII to XV
2. Sitting on the stretcher: movements VII and IX
3. Standing with the anterior portion of the thighs touching the stretcher: movements III, VI, VIII, X, XI, and XVI to XX

In this last position, be certain to prevent the occurrence of inappropriate joint movements that may compromise the measurement. Most of these subjects will score a 0 or 1 for these movements, with an occasional score of 2; some may not even be able to take the initial position for performing the movement (e.g., some shoulder movements).

The second situation requiring changes in Flexitest's methodology is when mobility is not measured for all 20 Flexitest movements due to a lack of time or the evaluator's lack of interest in obtaining all of the measurements. A common example of this is when the evaluator is at an elementary school or a sports club and has 200 or more children to teach. Because it takes three minutes to administer the Flexitest (although it tends to go faster in children, especially when they already know the test or have seen it administered to a classmate), it would take at least 10 hours to administer a complete Flexitest to all 200 students. Under these circumstances, it may be more convenient to use a condensed version of the Flexitest that has only a few movements. Although it is possible with sophisticated statistical techniques to select some of the most representative movements, it is probably more convenient for the evaluator to choose those measurements that are most appropriate for her purpose. For example, you would probably choose lower-limb movements to evaluate flexibility for soccer players, whereas a swimming coach would be more interested in measuring the mobility of his swimmers' ankles and shoulders. An evaluator who intends to use the Flexitest as one of a set of tests to assess health-related physical fitness might also find it impractical to use the full version, so the condensed model would be the best alternative. For instance, a five-movement Flexitest partial application, along with a broader array of tests, is being done on Brazilian Air Force personnel. But again, even though it is possible and quite appealing to calculate a condensed Flexindex (i.e., a global evaluation proportional to the number of movements performed), this would not be appropriate. Recently, after analyzing Flexitest data from our laboratory on over 2,600 subjects (Araújo, Oliveira, and Almeida 2002), we found that no combination of 6 to 10 movements provided a sufficiently low standard of error estimate to be of practical value. Considering these findings, those interested in using smaller or condensed versions of the Flexitest should be aware of the potential limitations of this approach. Whatever the situation, it is important to adhere to the intended philosophy and correct assessment of the maximum passive range of motion for the movement under evaluation. In this way, you will ensure that the measurement is as accurate and reliable as possible, enabling the data to be widely applicable and comparable.

Flexitest Practice

The Flexitest is a simple and effective method used to measure and assess flexibility for a wide variety of professional purposes. To assess a subject's flexibility, whether he is a student, patient, or athlete, there are two critical steps: *quality of movement execution* and *score quantification*. Once the Flexitest movement is properly performed, the ability to rate it accurately is critical. This issue is addressed in this chapter, which will permit you to practice and master the rating procedure.

In the following pages, you will find a total of 120 photos divided into two series. The first series presents four photos for each of the 20 movements (figures 5.1 to 5.20) in different arcs of movement. The second series—the last 40 photos—gives you a chance to assess your Flexitest scoring ability in two complete sequences of the 20 movements (figures 5.21 and 5.22). These series were designed to train you in two progressive steps: Initially, you have a chance to practice scoring each of the movements before moving on to the second series and testing your knowledge by evaluating two complete Flexitests. The photo of each movement is clearly identified (for example, 5.1a), and the correct score is presented at the end of the chapter (table 5.1 on page 109 for figures 5.1 to 5.20, table 5.2 on page 110 for figure 5.21, and table 5.3 on page 110 for figure 5.22) so you can check your ratings. These photos tend to replicate the view from the evaluator and observer positions seen in the evaluation maps. Notwithstanding, in order to provide you, as close as possible, the evaluator angle of view, the actual evaluator positioning in the photo is sometimes different than the one recommended in the description and in some selected cases, the evaluator is not appearing in the photo. The photos were chosen to reflect ordinary situations occurring when administering the Flexitest. Therefore, scores of 2 tend to be more common and scores of 0 or 4 may not be represented in a sample for a movement. Use the evaluation maps (see pages 52-71 in chapter 4) to score each of the photos. Write your rating scores on a separate sheet of paper, not in the book itself, so you can repeat the training as many times as you feel are necessary to improve your scoring skills.

> Additional Flexitest training materials—color digital images, slide shows, and videotapes—are also available for viewing and downloading at www.clinimex.com.br

Some movements naturally are more challenging to score, and in rare instances, it may be very difficult to decide which of two scores (e.g., 2 or 3) would be more suitable. Fortunately, these situations are the exception and not the rule, and with training, it will not be difficult for you to rate the photos quickly and accurately. Your margin of error should not exceed 1 point for each photo. If you make rating mistakes in more than 10% of the photos or make errors of more than 1 point for a single movement (e.g., if you award a score of 3 when the correct rating is a 1), your margin of error is too high. Return to chapter 4 and read again on pages 52-71 the descriptions of the movements, paying particular attention to the related comments and helpful hints. It may also be helpful to discuss your performance with a colleague who has experience scoring the Flexitest or has already completed the self-training in this chapter. Often, you will then be able to identify and rectify the mistakes that are causing a high margin of error.

After you finish self-training with the photos in this chapter, move on to practicing the evaluation of subjects—get your friends, family members, and colleagues to assist you in your learning. Always use the methodology description, the hints and comments, and the evaluation maps presented in chapter 4 as a guide. You will find that this seemingly subjective method is actually quite objective, easy to learn, easy to apply, and easy to interpret.

Movement I

Figure 5.1

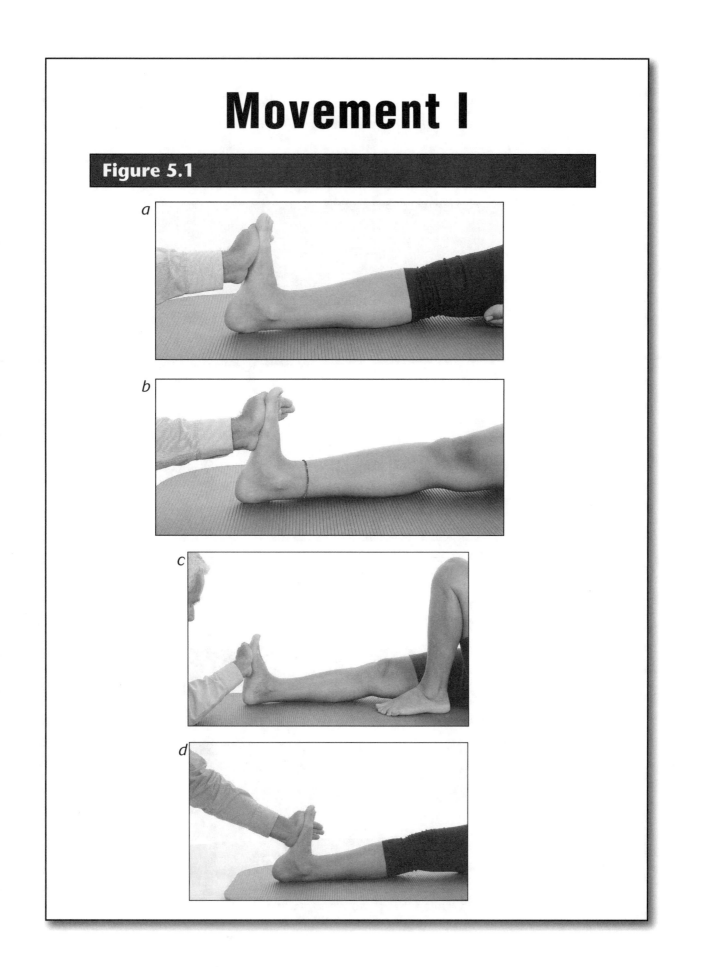

Movement II

Figure 5.2

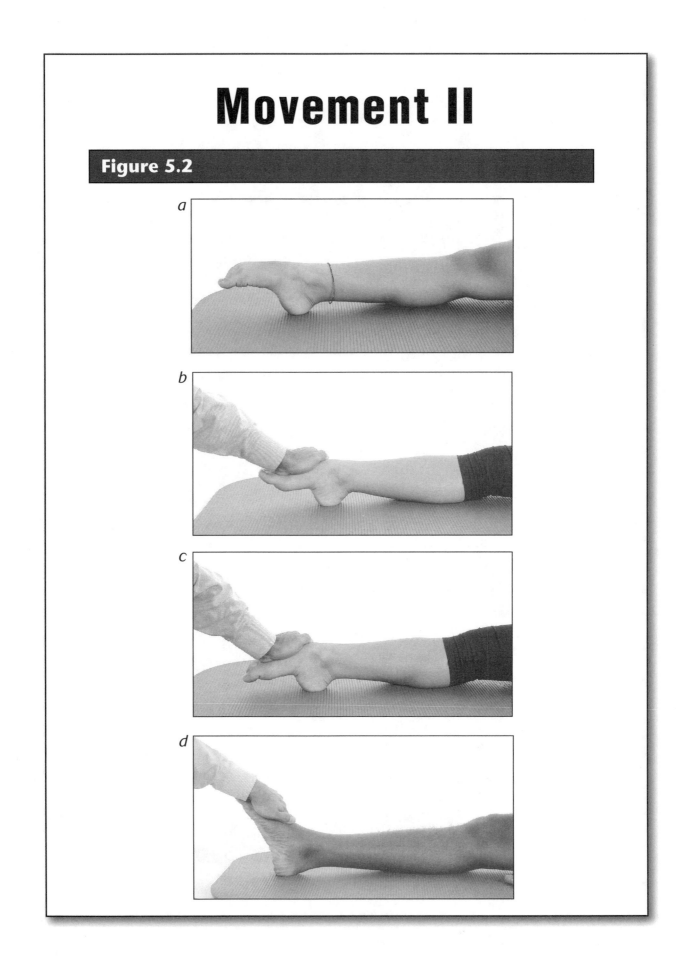

Movement III

Figure 5.3

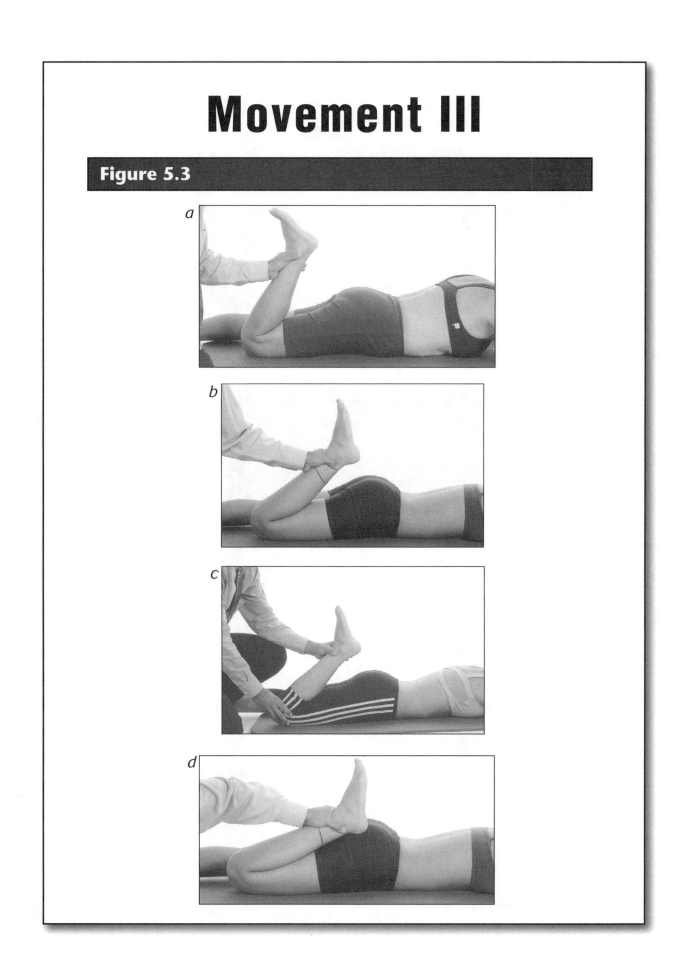

a

b

c

d

Movement IV

Figure 5.4

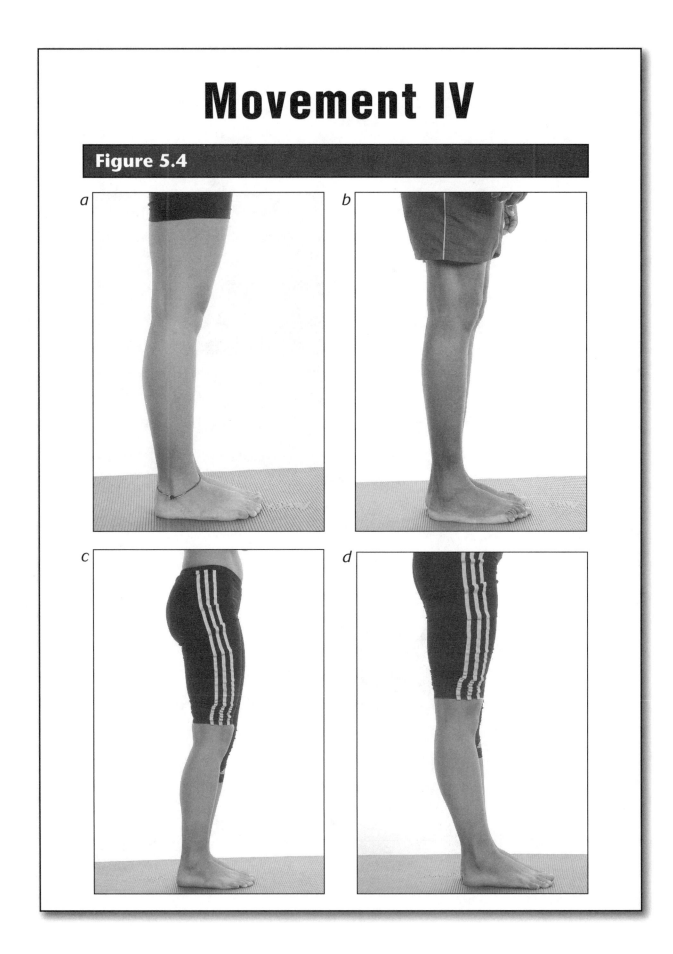

Movement V

Figure 5.5

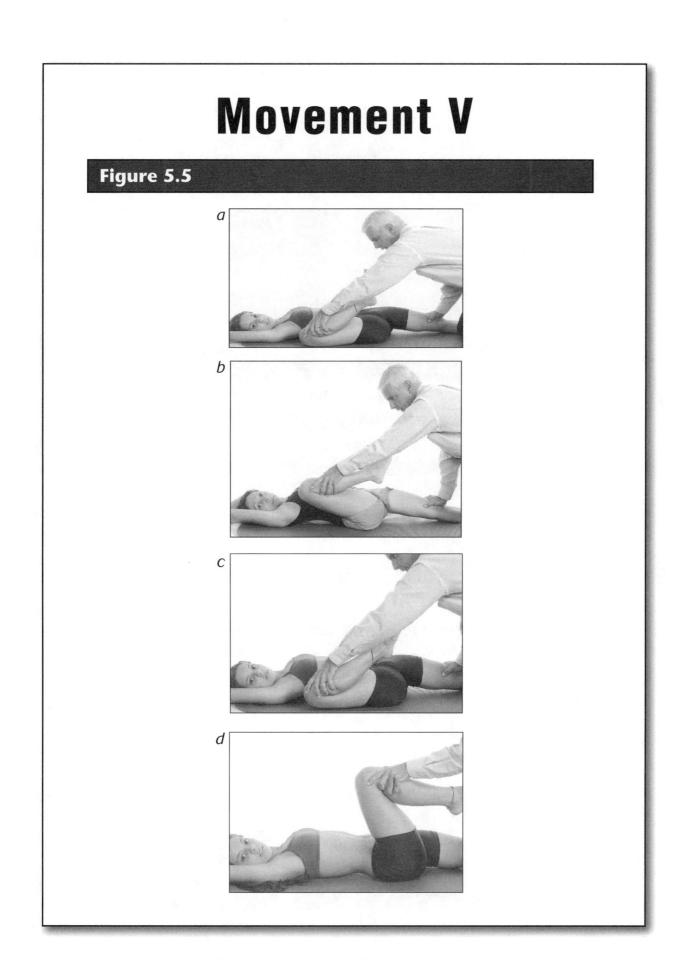

Movement VI

Figure 5.6

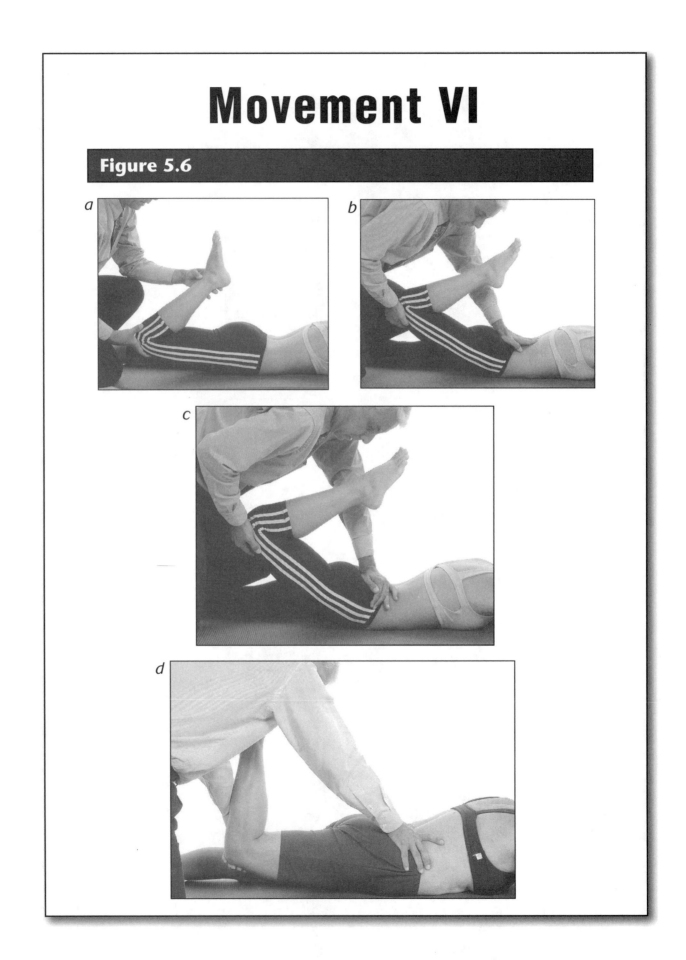

84

Movement VII

Figure 5.7

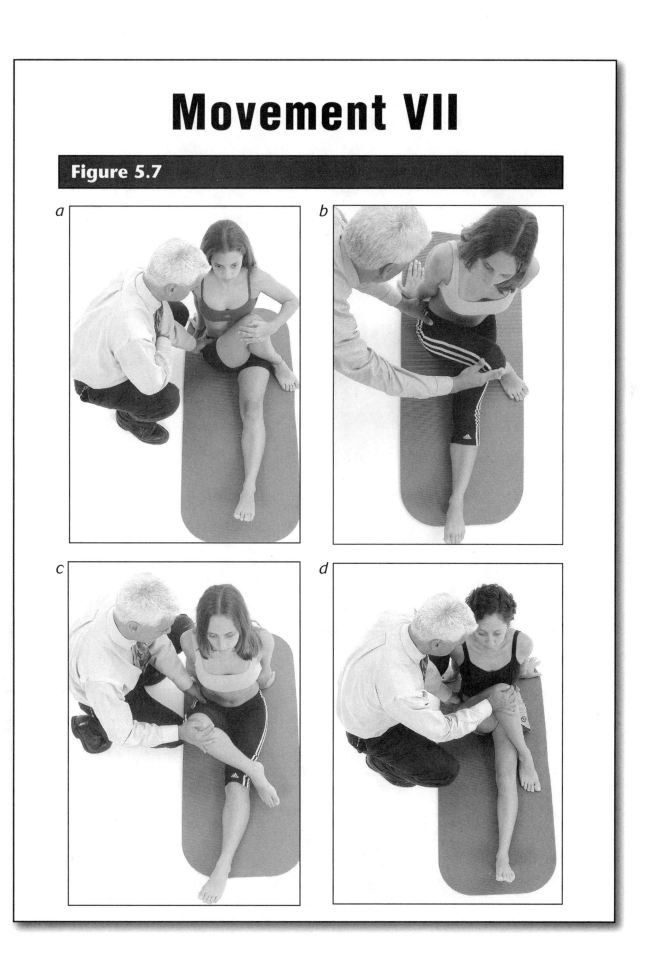

Movement VIII

Figure 5.8

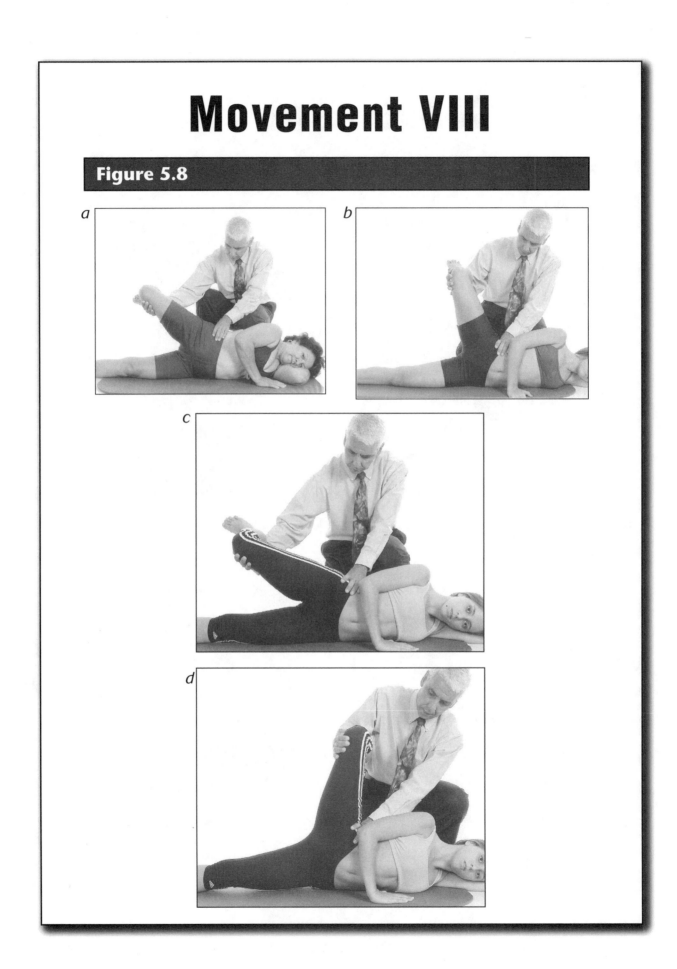

Movement IX

Figure 5.9

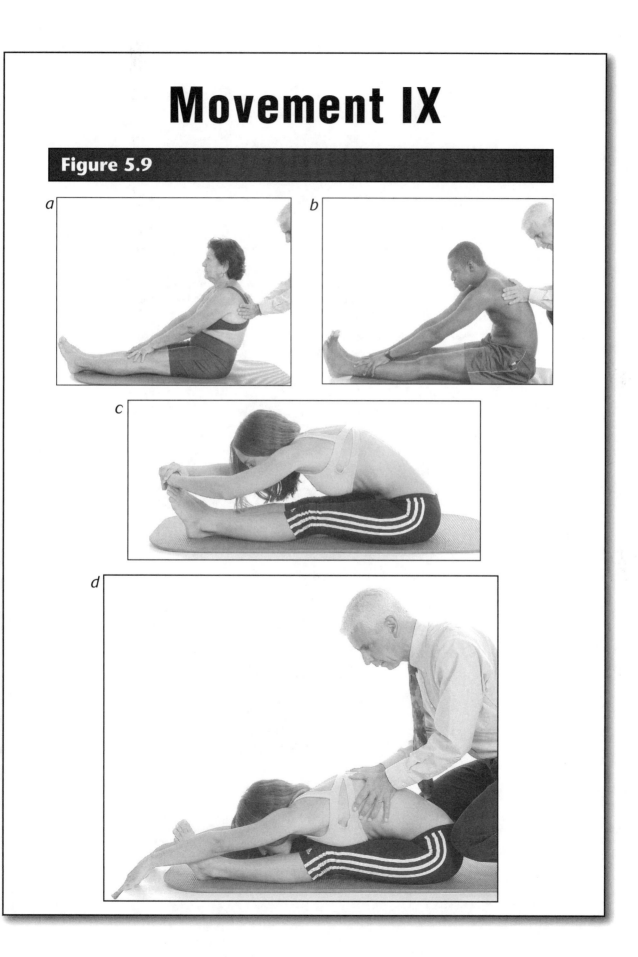

Movement X

Figure 5.10

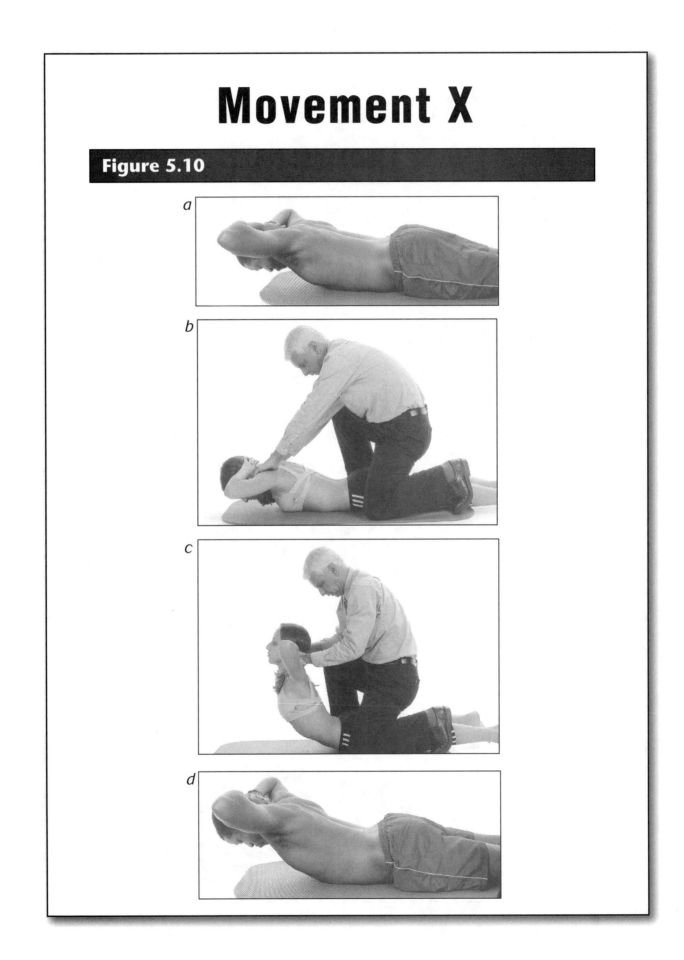

Movement XI

Figure 5.11

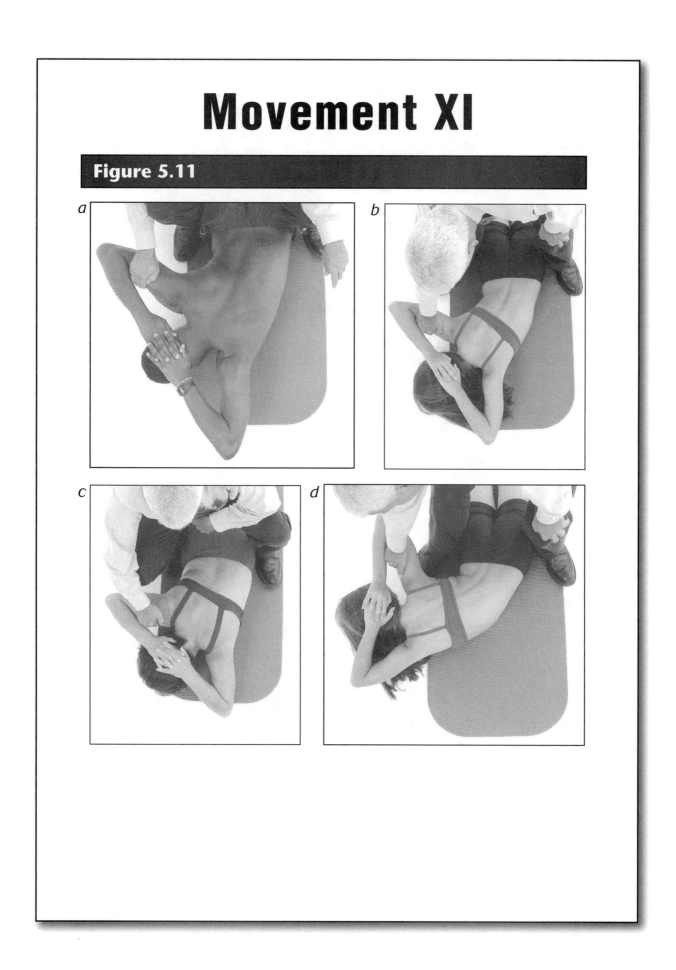

Movement XII

Figure 5.12

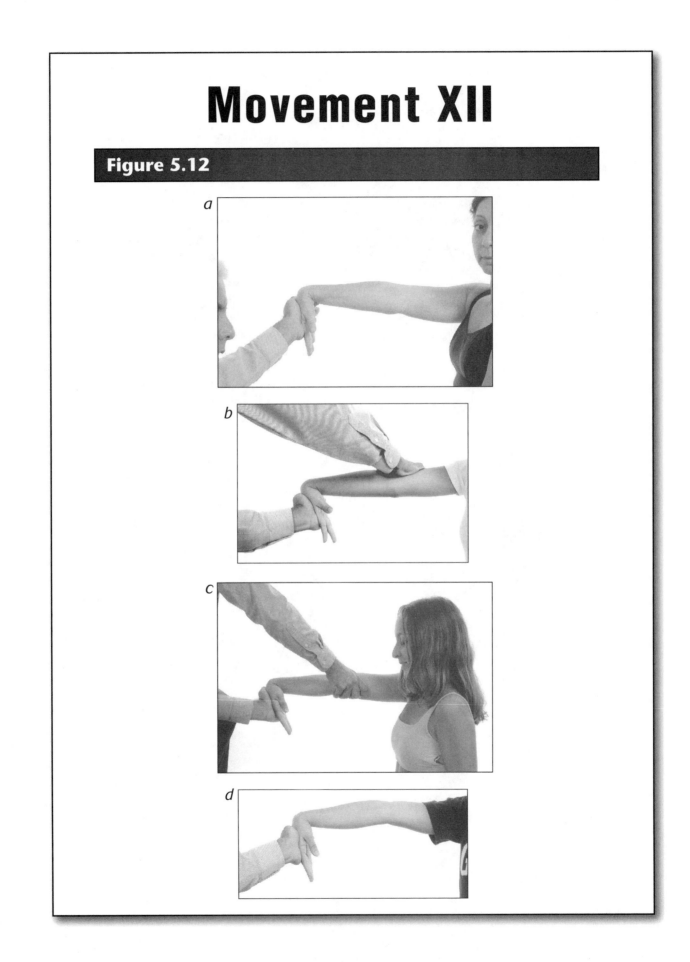

Movement XIII

Figure 5.13

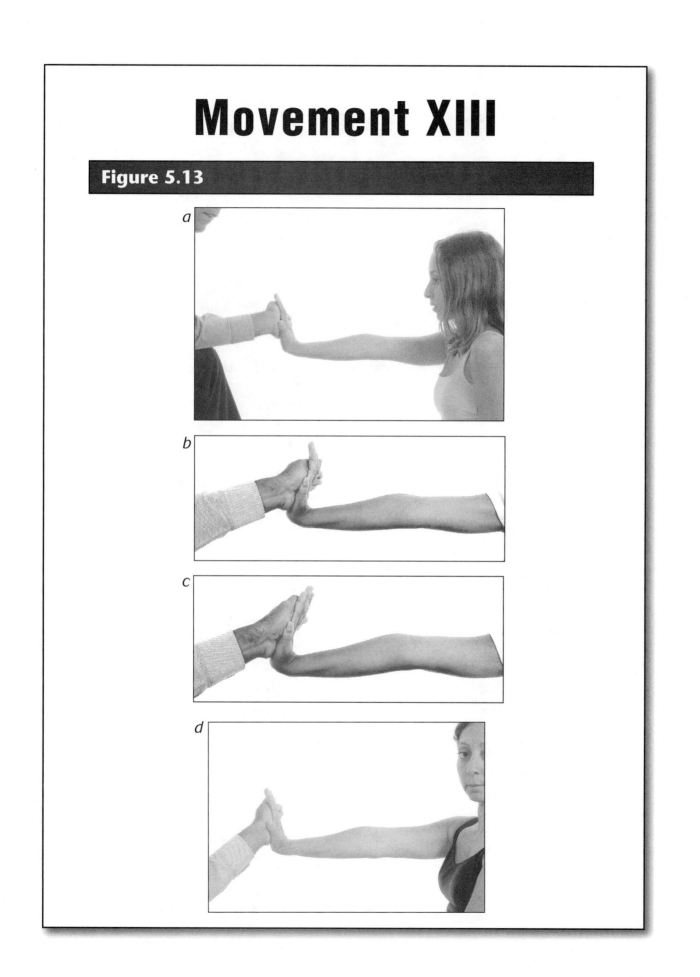

Movement XIV

Figure 5.14

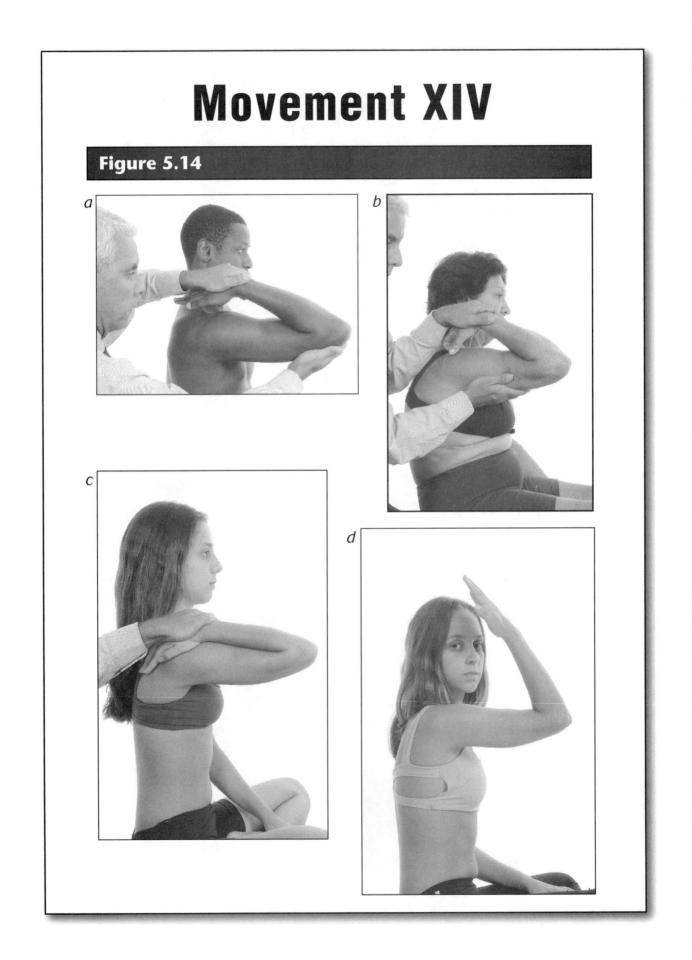

Movement XV

Figure 5.15

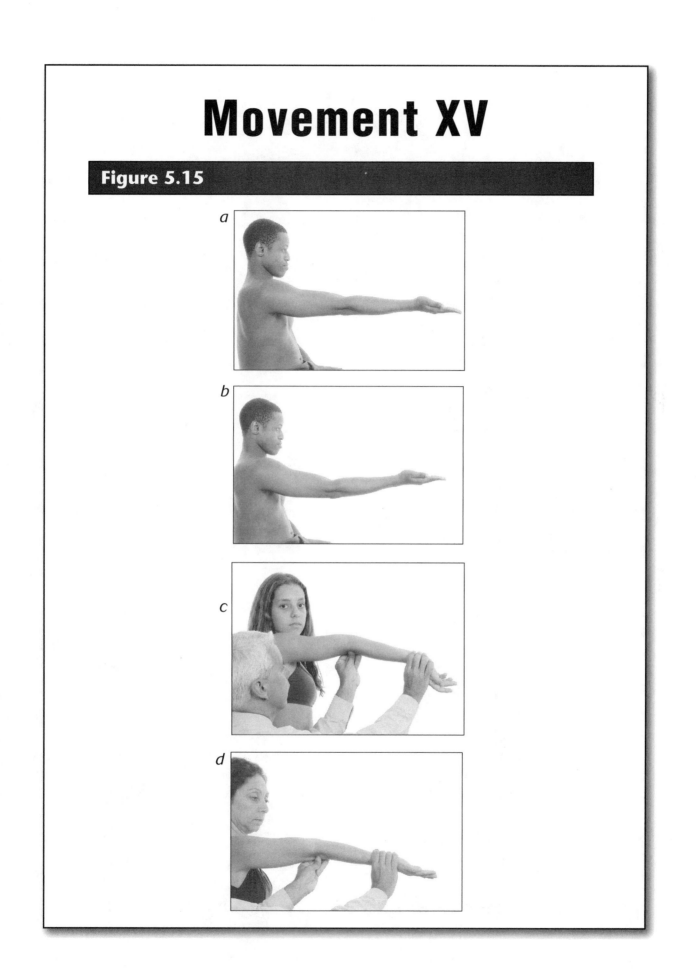

Movement XVI

Figure 5.16

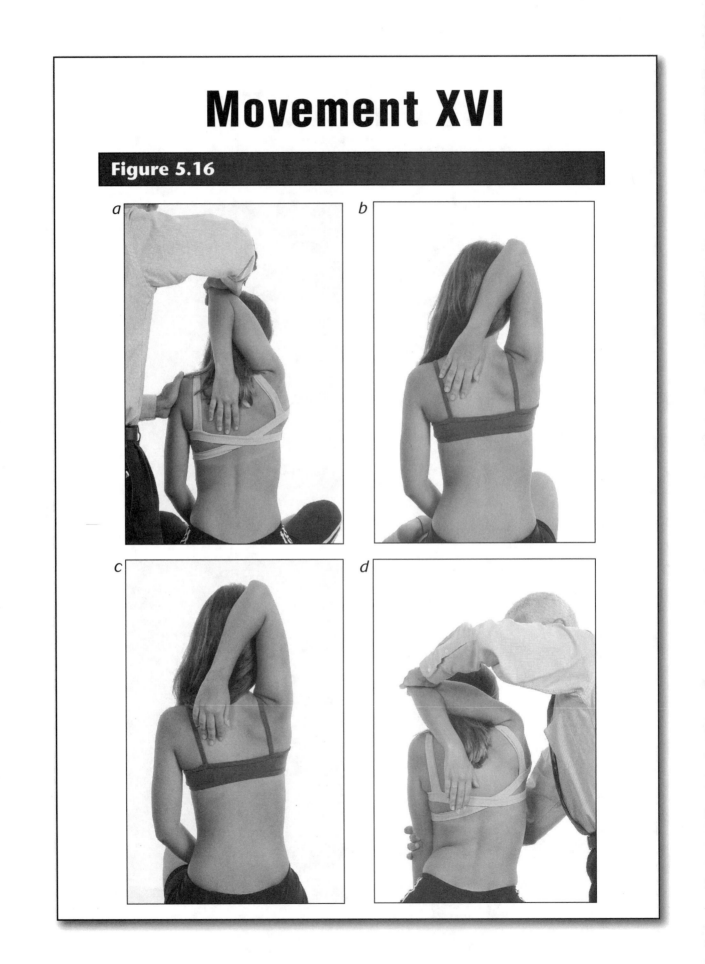

Movement XVII

Figure 5.17

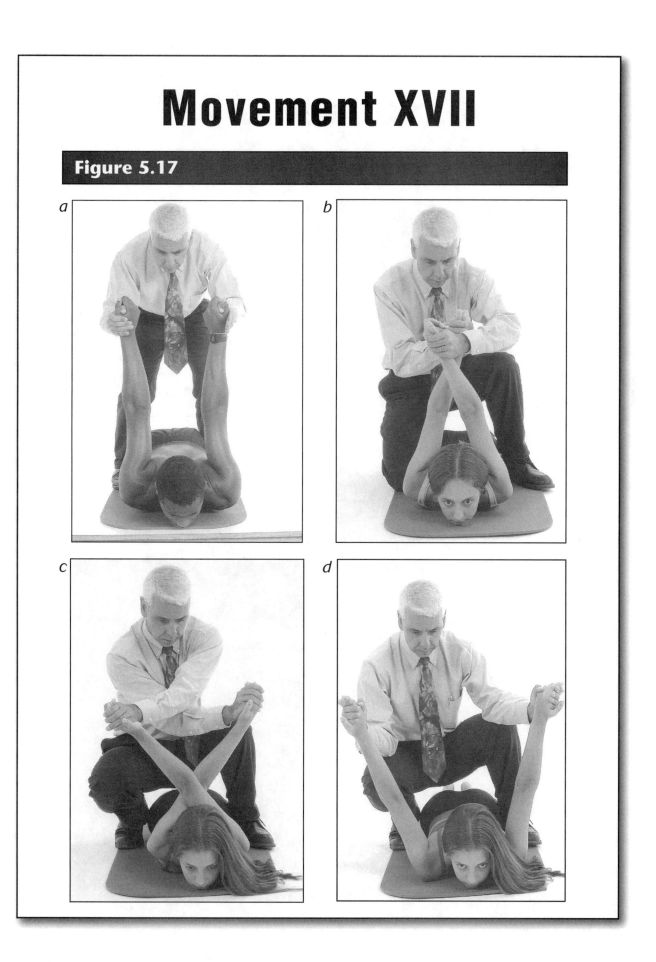

Movement XVIII

Figure 5.18

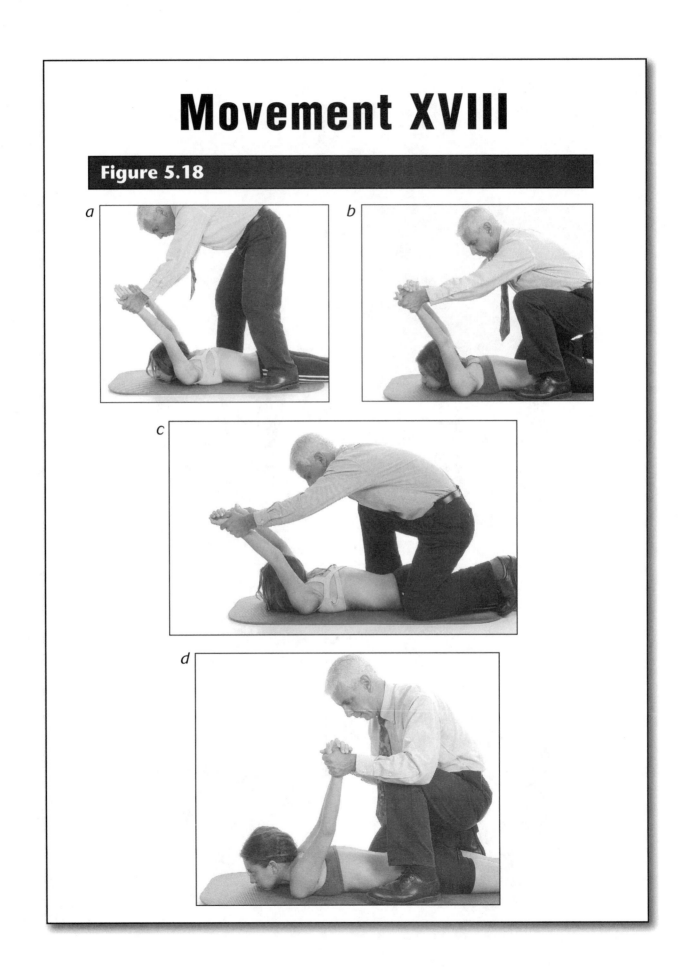

Movement XIX

Figure 5.19

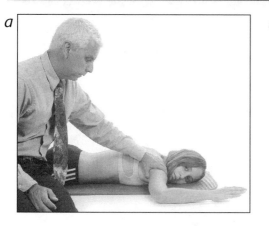

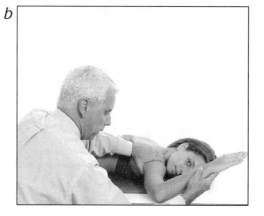

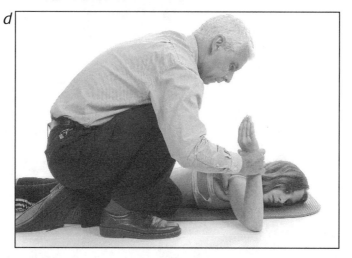

Movement XX

Figure 5.20

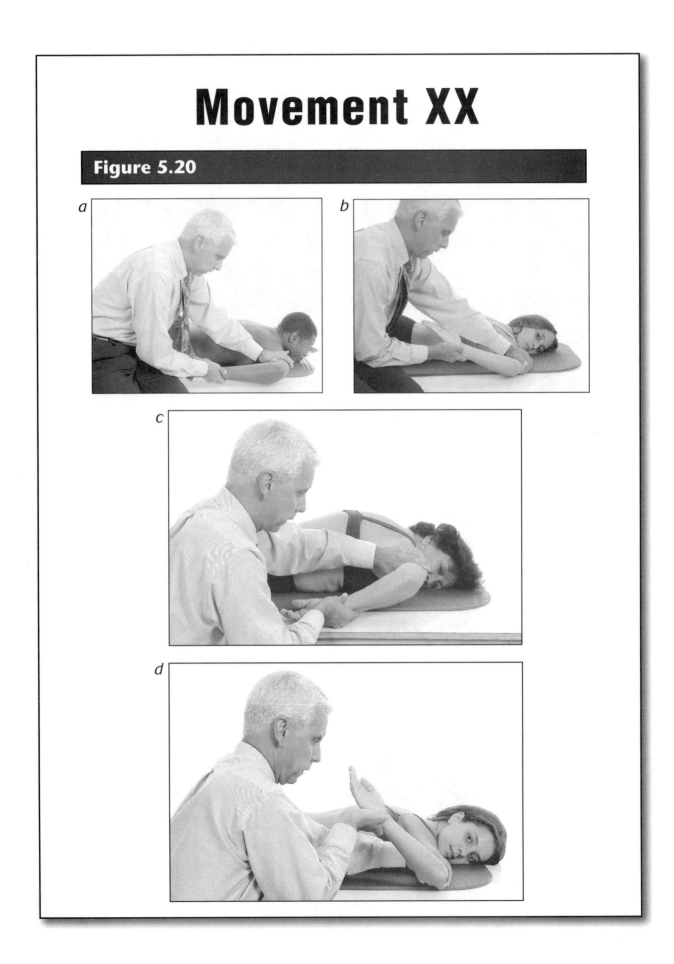

Movement Sequence 1

Figure 5.21

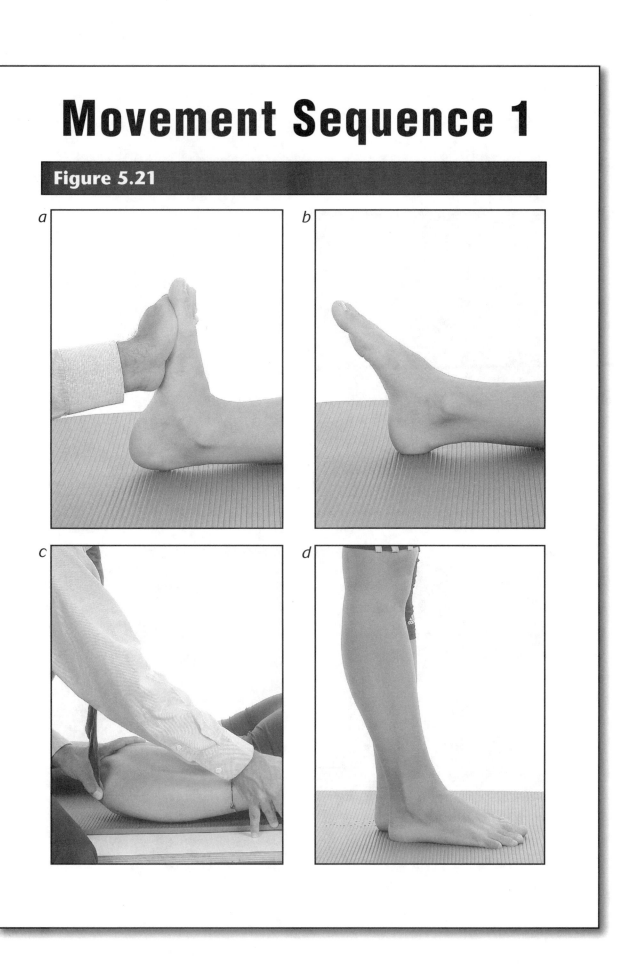

Movement Sequence 1

Figure 5.21 *(continued)*

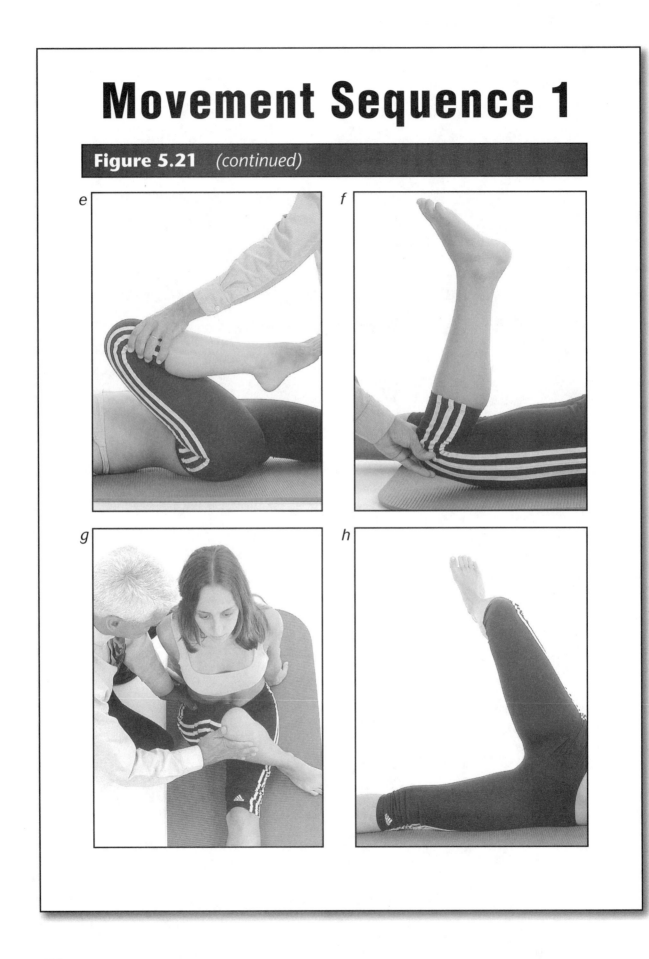

Movement Sequence 1

Figure 5.21 *(continued)*

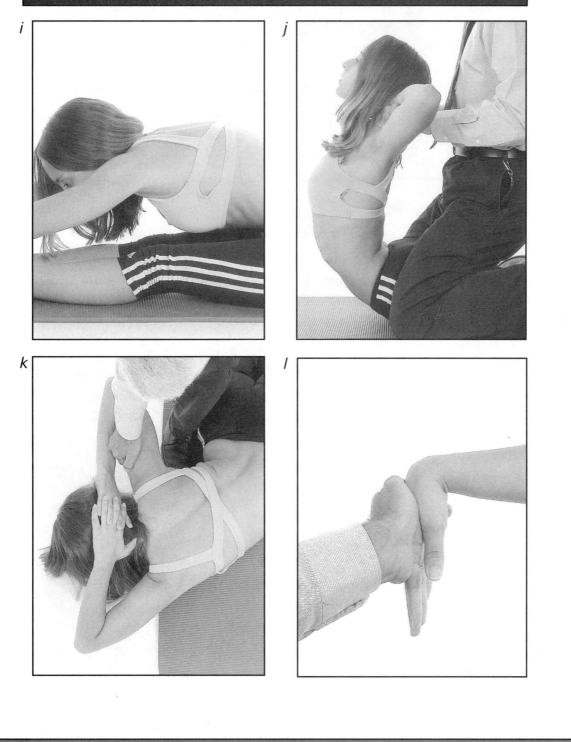

Movement Sequence 1

Figure 5.21 *(continued)*

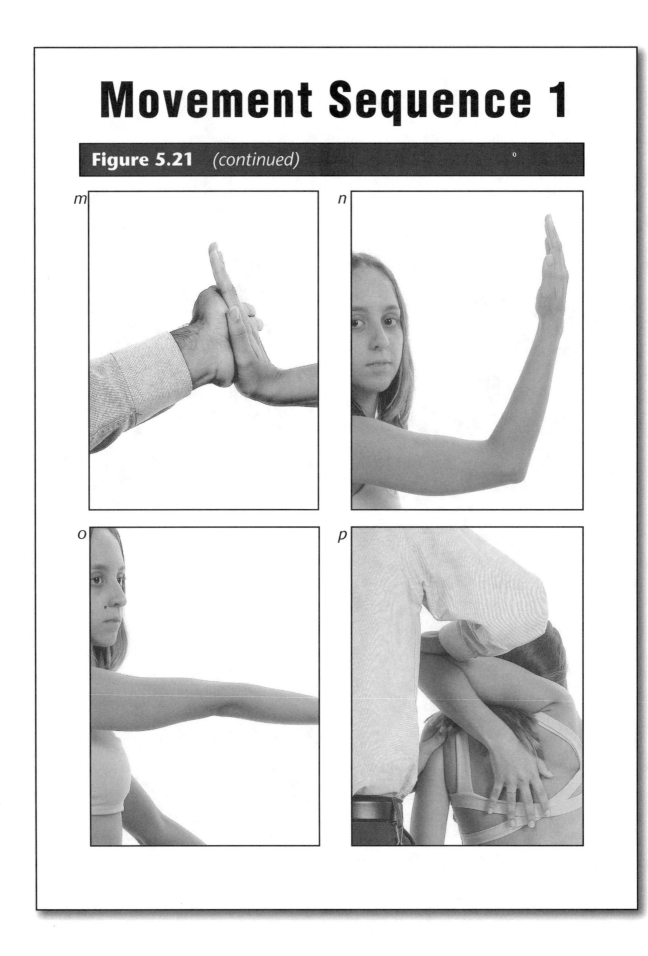

Movement Sequence 1

Figure 5.21 *(continued)*

q

r

s
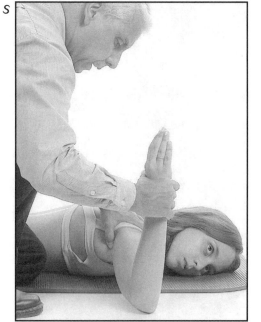

t
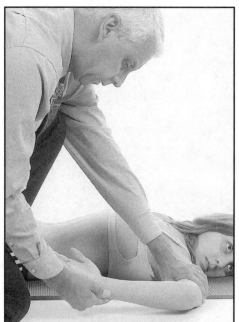

Movement Sequence 2

Figure 5.22

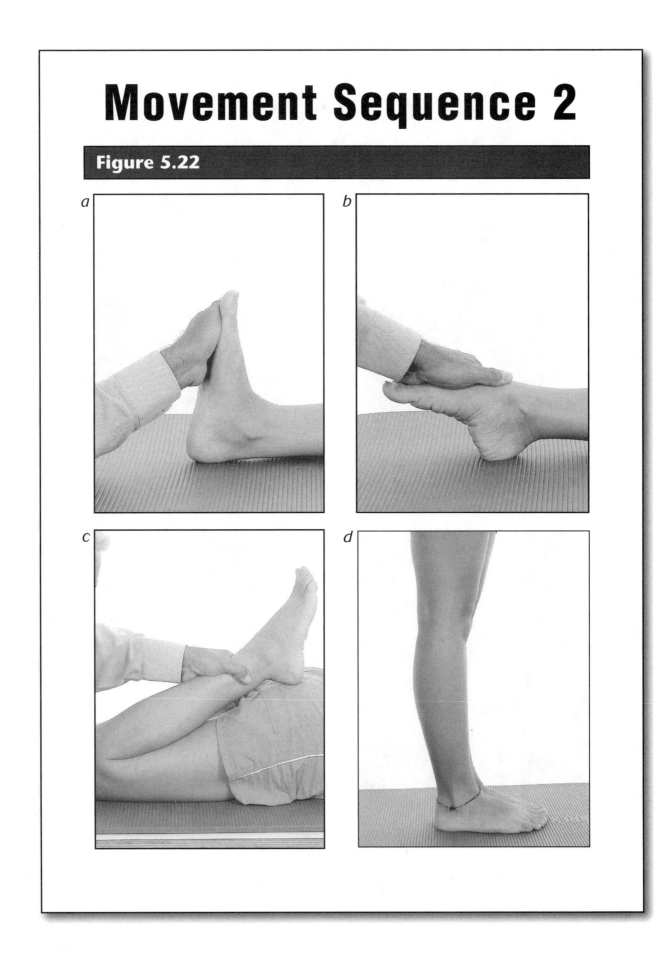

Movement Sequence 2

Figure 5.22 *(continued)*

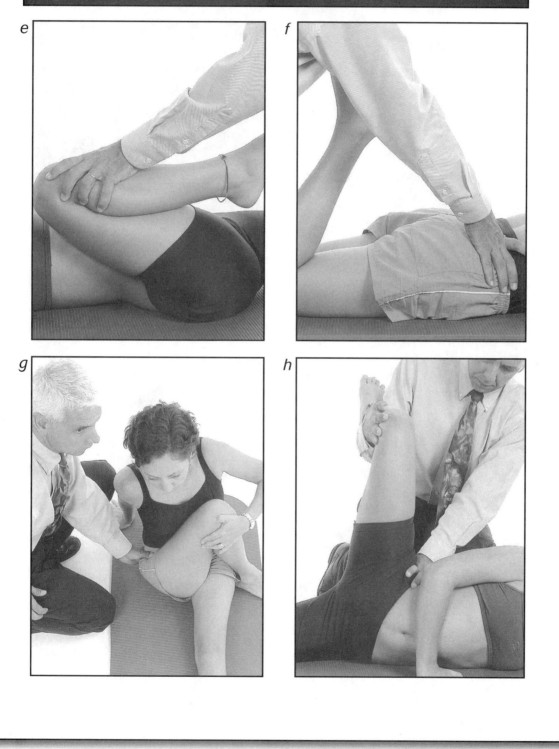

Movement Sequence 2

Figure 5.22 *(continued)*

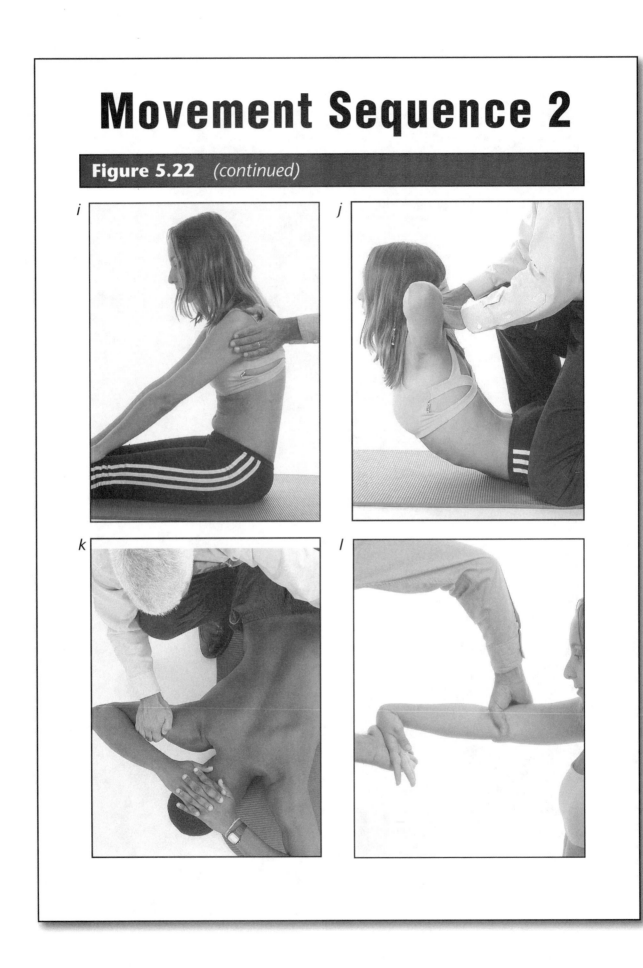

Movement Sequence 2

Figure 5.22 *(continued)*

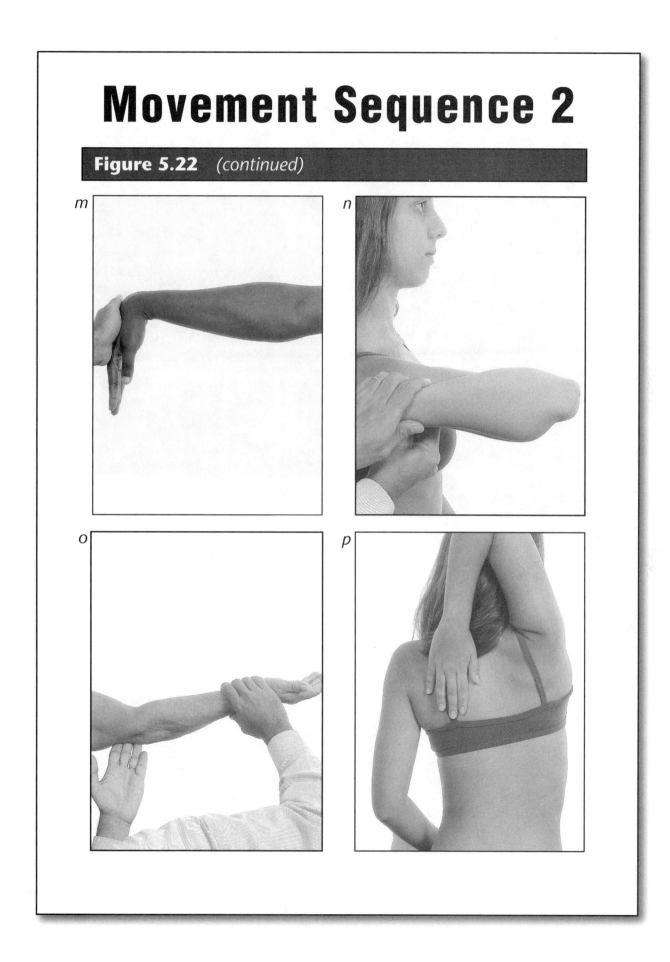

Movement Sequence 2

Figure 5.22 *(continued)*

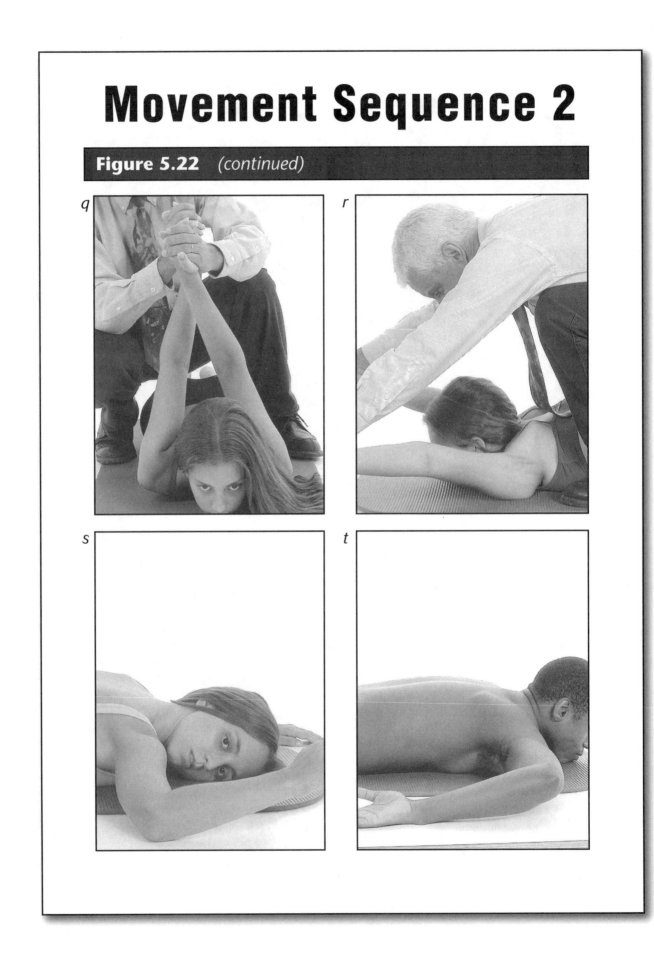

Table 5.1				Scoring Guide for Figures 5.1 Through 5.20	
Movement I	**Score**	**Movement II**	**Score**	**Movement III**	**Score**
Figure 5.1a	2	Figure 5.2a	2	Figure 5.3a	1
Figure 5.1b	2	Figure 5.2b	3	Figure 5.3b	2
Figure 5.1c	2	Figure 5.2c	3	Figure 5.3c	2
Figure 5.1d	3	Figure 5.2d	1	Figure 5.3d	3
Movement IV	**Score**	**Movement V**	**Score**	**Movement VI**	**Score**
Figure 5.4a	2	Figure 5.5a	4	Figure 5.6a	1
Figure 5.4b	2	Figure 5.5b	3	Figure 5.6b	2
Figure 5.4c	1	Figure 5.5c	4	Figure 5.6c	3
Figure 5.4d	3	Figure 5.5d	1	Figure 5.6d	0
Movement VII	**Score**	**Movement VIII**	**Score**	**Movement IX**	**Score**
Figure 5.7a	4	Figure 5.8a	2	Figure 5.9a	0
Figure 5.7b	3	Figure 5.8b	3	Figure 5.9b	2
Figure 5.7c	0	Figure 5.8c	1	Figure 5.9c	3
Figure 5.7d	3	Figure 5.8d	4	Figure 5.9d	4
Movement X	**Score**	**Movement XI**	**Score**	**Movement XII**	**Score**
Figure 5.10a	0	Figure 5.11a	2	Figure 5.12a	2
Figure 5.10b	0	Figure 5.11b	2	Figure 5.12b	3
Figure 5.10c	3	Figure 5.11c	1	Figure 5.12c	3
Figure 5.10d	1	Figure 5.11d	4	Figure 5.12d	2
Movement XIII	**Score**	**Movement XIV**	**Score**	**Movement XV**	**Score**
Figure 5.13a	1	Figure 5.14a	2	Figure 5.15a	2
Figure 5.13b	2	Figure 5.14b	2	Figure 5.15b	1
Figure 5.13c	2	Figure 5.14c	3	Figure 5.15c	3
Figure 5.13d	1	Figure 5.14d	1	Figure 5.15d	2
Movement XVI	**Score**	**Movement XVII**	**Score**	**Movement XVIII**	**Score**
Figure 5.16a	2	Figure 5.17a	2	Figure 5.18a	2
Figure 5.16b	1	Figure 5.17b	3	Figure 5.18b	2
Figure 5.16c	1	Figure 5.17c	4	Figure 5.18c	3
Figure 5.16d	4	Figure 5.17d	1	Figure 5.18d	1
Movement XIX	**Score**	**Movement XX**	**Score**		
Figure 5.19a	0	Figure 5.20a	1		
Figure 5.19b	1	Figure 5.20b	2		
Figure 5.19c	3	Figure 5.20c	0		
Figure 5.19d	4	Figure 5.20d	4		

Table 5.2 — Scoring Guide for Figure 5.21

Movement I	Score	Movement II	Score	Movement III	Score
Figure 5.21*a*	2	Figure 5.21*b*	1	Figure 5.21*c*	4
Movement IV	**Score**	**Movement V**	**Score**	**Movement VI**	**Score**
Figure 5.21*d*	3	Figure 5.21*e*	2	Figure 5.21*f*	0
Movement VII	**Score**	**Movement VIII**	**Score**	**Movement IX**	**Score**
Figure 5.21*g*	2	Figure 5.21*h*	2	Figure 5.21*i*	3
Movement X	**Score**	**Movement XI**	**Score**	**Movement XII**	**Score**
Figure 5.21*j*	4	Figure 5.21*k*	4	Figure 5.21*l*	2
Movement XIII	**Score**	**Movement XIV**	**Score**	**Movement XV**	**Score**
Figure 5.21*m*	0	Figure 5.21*n*	0	Figure 5.21*o*	2
Movement XVI	**Score**	**Movement XVII**	**Score**	**Movement XVIII**	**Score**
Figure 5.21*p*	4	Figure 5.21*q*	2	Figure 5.21*r*	3
Movement XIX	**Score**	**Movement XX**	**Score**		
Figure 5.21*s*	3	Figure 5.21*t*	1		

Table 5.3 — Scoring Guide for Figure 5.22

Movement I	Score	Movement II	Score	Movement III	Score
Figure 5.22*a*	2	Figure 5.22*b*	2	Figure 5.22*c*	3
Movement IV	**Score**	**Movement V**	**Score**	**Movement VI**	**Score**
Figure 5.22*d*	2	Figure 5.22*e*	3	Figure 5.22*f*	0
Movement VII	**Score**	**Movement VIII**	**Score**	**Movement IX**	**Score**
Figure 5.22*g*	4	Figure 5.22*h*	4	Figure 5.22*i*	1
Movement X	**Score**	**Movement XI**	**Score**	**Movement XII**	**Score**
Figure 5.22*j*	3	Figure 5.22*k*	2	Figure 5.22*l*	3
Movement XIII	**Score**	**Movement XIV**	**Score**	**Movement XV**	**Score**
Figure 5.22*m*	1	Figure 5.22*n*	4	Figure 5.22*o*	0
Movement XVI	**Score**	**Movement XVII**	**Score**	**Movement XVIII**	**Score**
Figure 5.22*p*	2	Figure 5.22*q*	3	Figure 5.22*r*	4
Movement XIX	**Score**	**Movement XX**	**Score**		
Figure 5.22*s*	1	Figure 5.22*t*	0		

Flexitest Analysis

By knowing that a male student's height is 6 feet, one can relate it to referential standards and conclude whether this student's height is below, near, or above his male peers' typical height at that age. The same line of reasoning goes for Flexitest results: Data are compared to the values expected for the subject's age and gender to determine the subject's relative flexibility. After learning, practicing, and applying Flexitest, one must consider the basic question of how to move from measuring to evaluating. This chapter introduces major descriptive and inferential statistical approaches for a scientifically based evaluation of Flexitest results. In this chapter, we consider four possible scenarios for Flexitest analysis. It may be necessary to do the following:

1. Analyze a subject's result as a function of what was expected for the corresponding age group, gender, or sport modality.
2. Describe and summarize data from a group of subjects.
3. Compare results of two or more groups of subjects.
4. Compare results before and after intervention (e.g., a school gym class or a specific fitness training program).

Because Flexitest data may be analyzed collectively with Flexindex, by movement, by joint, and by homogeneity profile, techniques are presented for each of these approaches. Although this book is not intended to be a text on statistics, some statistical terminology and concepts are addressed to make easier the use of resources for interpreting Flexitest results.

Preliminary Statistical Considerations

Flexitest uses whole numbers from 0 to 4 to measure the passive joint mobility of each movement. Because there are no intermediate or fractional values, this variable can be considered as mathematically *discontinuous*, even though the variable being measured is intrinsically continuous.

With respect to measurement scales, Flexitest scores may be ranked according to categories that have intervals arbitrarily defined as equals, but do not have a true zero. Therefore, it is possible to classify each Flexitest movement on an *interval scale* of measurement, because a score of zero does not designate a complete lack of mobility. On the other hand, evaluators using a more conservative statistical approach might classify Flexitest results on the *ordinal scale* of measurement. Statistical techniques tend to be more complicated and less powerful for a lower hierarchical level of the measurement scale, so we suggest the interval scale be employed whenever possible, especially when analyses deal with rather large numbers of subjects (i.e., more than 30).

To analyze collected data, we use descriptive statistical techniques to summarize them as *central tendency* and *variability indicators,* the former type representing the group and the latter a number or range indicating the extent of deviation from this central value. Central tendency is best represented by the *median* score, although the *mode* (the most frequent score) may also be used. To determine the median, also called the

50th percentile, it is necessary only to identify the score that is in the middle of all the scores listed from the lowest to the highest. Calculation and interpretation of a *mean* score should be carefully made, because there may be some distortion with a scale that has only five possible scores, especially when the sample size is small. *Variability* may be defined as the limits of the extreme results for a movement in a given sample or as the proportion of subjects with scores different from the mode. Again, making use of more advanced variability analysis techniques, such as *standard deviation* and *standard error* of the mean, presents the same conceptual limitations as use of the mean and, in principle, should not be included in the analysis of Flexitest data for a single movement.

Results may also simply be listed from the lowest to the highest value; when they are presented in this way, it is appropriate to indicate the relative rank of the subject as a proportion of the whole sample. For this purpose, one can use *quartiles* (divisions that each contain results for 25% of the total population), *deciles* (divisions of 10%), and, more often, *percentiles* (1%). For example, in a sample of 100 students ranking individual Flexindex scores from the bottom up, a student with a 25th percentile (P_{25}) or 1st quartile (Q_1) has an overall flexibility higher than that of 25% of his peers and lower than that of the other 75%. This technique is frequently used to monitor children's height and weight as compared to growth curves and is especially useful for interpreting a subject's results for different movements and the whole Flexitest set as a function of age and gender.

Hypothesis tests compare results between two or more groups of subjects, or between different measurements in the same group of subjects, to allow the researcher to infer if extrapolated differences are likely to be actual distinctions or primarily due to chance. In this technique (Chi-square statistics to compare results from a single movement), a *contingency table* is mounted and *chi-square statistics* are calculated. It is then possible to determine whether the distributions of two sets of Flexitest results are significantly different from one another. It may be necessary to group results from extreme scores, because the information cannot be used when any of the table cells are empty (or present less than 5% of the counts), i.e., when the frequency is zero.

Another evaluative possibility permitted by the ordinal nature of Flexitest data is the application of the *Kolmogorov-Smirnov test*, which identifies critical differences in each score when two samples are compared. However, this procedure may be too conservative in that it is very difficult to find significant differences, thus limiting its practical use. When three or more groups are compared or the same group is repeatedly tested, chi-square statistics can be calculated by post hoc comparison techniques used to identify distributions that differ from one another. With large and normally distributed samples, and within the same theoretical limitations and restrictions, parametric statistics—including t-tests and analyses of variance—may be applied.

If your purpose is to determine the reliability of two sets of data for the same movement (i.e., to check your execution technique), it is not advisable to determine the *proportion of agreement*, although *kappa statistics* can be used because they correct for coincidences. Neither is using the *Pearson product-moment correlation* recommended, because it measures only the association between two series, which may be high even when significant differences between the two measurement sets exist. Instead, it is preferable to calculate the *intraclass correlation coefficient*, which is obtained from an analysis of variance and estimates true variability in two or more evaluators, thus more realistically reflecting the desired index.

Furthermore, to identify any existing association between Flexitest results and any other variable, *correlation* and *regression analysis* techniques are used. There are a number of books on applied statistics that detail the procedures we suggest here, as well as commercial statistical software packages and even electronic spreadsheets that may be used for Flexitest data analysis.

Complete Flexitest Set Analysis (Flexindex)

One of the most common uses for Flexitest is to assess the joint mobility of an individual, whether it is a student, patient, or athlete. Once the 20 movements are rated, they must be evaluated so practical meaning can be established for those results. One of the Flexitest's major advantages is that a score of 2 for one movement has the same meaning for all movements in terms of the relative magnitude of the range of motion. For example, when a subject scores a 2 for both hip extension and trunk extension, it means that his flexibility level is similar for both movements. The mode of Flexitest results is 2 for all young adults, and the measurement scale is ordinal (hierarchically superior to interval or ratio scales). Therefore one can add values to calculate partial indexes per joint or segment or even an overall flexibility index, Flexindex, which is obtained by adding the scores of all 20 movements.

Analyzing Individual Data

Using Flexindex, an individual's overall flexibility can easily be compared to percentile curves for age and gender. By using the same percentiles as in Children´s Growth Curves recently adopted by the U.S. Centers for Disease Control and Prevention, we have determined Flexitest curves for percentiles 3, 10, 25, 40, 60, 75, 90, and 97 for each age group and gender. The plotting of an individual's results, whether from one Flexitest or a number over the years, provides a simple way to visualize a subject's

results and interpret them in comparison to the peer group. Some expressions may be applied to each range, although numerical data provide more accurate pictures than do verbal expressions (table 6.1).

As mathematically defined by age- and gender-related percentiles, half of all people rank in the intermediate levels, i.e., between P_{25} and P_{75}, so that range is the expected flexibility standard, especially for subjects whose Flexindex results are close to the median, or P_{50}. Those with P_3 to P_{25}—very low and low—overall flexibility results would probably benefit if they undertook hard and specific training on it. Results at each end of the scale, under P_3 or over P_{97}, are somewhat uncommon and statistically abnormal (out of the 95% range of normality), and are often associated with clinical conditions of hypo- and hypermobility.

Age and Gender Reference Curves   Age and gender Flexitest reference curves will aid you in interpreting data from individual male and female subjects between 5 and 88 years of age. The curves were derived from our database of a total of 1,847 men and 1,269 women measured by evaluators highly skilled in using the Flexitest technique (a large majority of them were measured by the author) (figures 6.1 and 6.2). Most children underwent the Flexitest in their school's physical education classes or at fitness clubs. The 16- to 25-year-old age group was drawn primarily from college students, with the majority of them being physical education majors. Data from the age groups of subjects older than 25 years came from clients seen at our Sports and Exercise Medicine Clinic either to undergo a morphofunctional evaluation or to join a medically supervised exercise program. Youngsters or adults who participated in competitive sports were excluded from this analysis.

The complete database is further divided into 20 age groups for each gender, based on consideration of statistical, technical, and operational aspects. Inclusion criteria conformed to chronological age, not biological age (as reflected by maturity or other functional features). Our first consideration in the division of

Table 6.1	Evaluation Expression in Relation to Flexindex Percentiles
Percentile range	**Expression**
Under P_3	Extremely low
Between P_3 and P_{10}	Very low
Between P_{11} and P_{25}	Low
Between P_{26} and P_{40}	Lower average
Between P_{41} and P_{60}	Average
Between P_{61} and P_{75}	Upper average
Between P_{76} and P_{90}	High
Between P_{91} and P_{97}	Very high
Over P_{97}	Extremely high

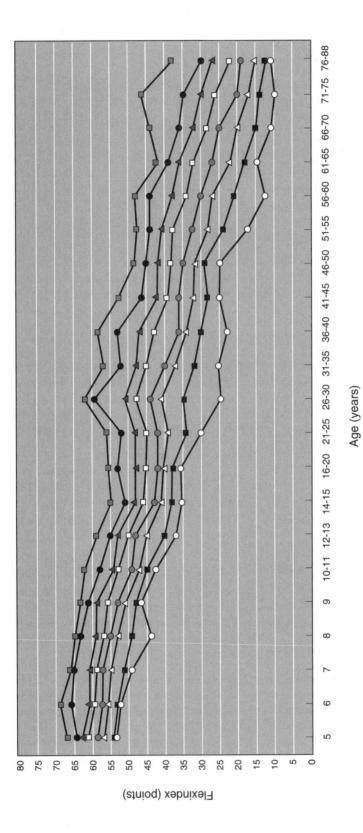

Figure 6.1 Flexindex—male percentile curves.

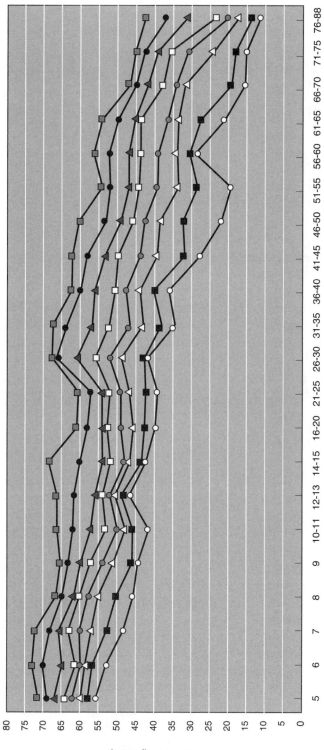

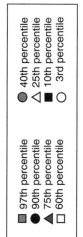

Figure 6.2 Flexindex—female percentile curves.

115

age groups was to prevent significant Flexindex variations between consecutive ages from being missed. Flexibility tends to decrease with aging, and women are typically more flexible than men of a similar age. However, the magnitude of loss varies considerably from one age to another, with greater decreases seen during the early years than in mature adults. Therefore, we determined percentiles for each age from 5 to 88 years (see tables 6.2 and 6.3). At this point the magnitude of flexibility reduction slows, and we began to group together values for subjects within two years of age. Age groups at five-year intervals were formed starting at age 16. This age corresponds to the end of adolescence and the beginning of adult life; consequently, there is a slowing of the flexibility decrement, reducing the need for frequent group divisions. This grouping allows for a larger sample, and thus more reliable data. Five-year age groupings were used up to the age group of 60-65, after which the interval increased to 10 years due to the limited number of subjects in our data bank. All age groups contained a minimum of 30 cases, with the exception of the female age group 75+, which was a 12-subject sample.

> Individual Flexindex scores (the sum of all 20 movement scores) should always be analyzed against age and gender reference norms.

Detailed observation of the percentile curves as a function of age for both genders reflects an atypical increase of flexibility in the variability (range) for the age group 26-30 years. We hypothesize that this is the result of a sampling problem—the subject pool for this age group was relatively small and may be somewhat biased by mixing subjects with quite different patterns of regular physical activity as compared with those in the other sampled age groups. The data should not be interpreted to mean that flexibility effectively increased over this period of life. With this caveat noted, we made no trend adjustments.

Comparative analysis of individual Flexindex results with reference data allows goals to be defined and the impact of intervention or training to be assessed. For instance, an individual with a very low Flexindex score may,

after three months of specific training with stretching exercises, move to a low-rank score, showing the partial success of the intervention and the need for the subject to continue training to reach higher flexibility levels.

Sport-Related Aspects   Collecting data from elite athletes is a methodologically complex issue. It is difficult, if not impossible, to standardize measurement conditions for athletes, because doing so may partially compromise data reliability. Several factors complicate the collection of reliable data from elite athletes:

▼ Limited availability of athletes and coaches for this type of assessment

▼ High variability caused by evaluation being performed in various phases of training

▼ Variability in the amount of time elapsed since the last training session

▼ Presence of acute or chronic lesions or partial orthopedic limitations (leading to underestimation of the actual maximum range of movement)

▼ Frequent use of medications that can affect mobility (for instance, anti-inflammatory and muscle-relaxing drugs)

▼ Above all, the athlete's degree of excellence or proficiency as compared to their peers at the time of measurement

After collecting Flexitest data from hundreds of high-level competitive athletes (many of them of international or Olympic caliber, especially in women's beach volleyball) of both genders between 15 and 35 years of age and involved in different sport modalities, we saw that, as a whole, athletes have overall flexibility levels similar to those of nonathletes. Figures 6.3 and 6.4 reflect these Flexindex scores.

One would expect athletes to devote more time to stretching exercises than nonathletes, and that may explain their higher amplitude in movements such as trunk flexion and hip adduction, which are commonly included in training or competition warm-up series. Interestingly, shoulder mobility tended to be lower in athletes than in nonathletes, with the exception of athletes involved in all swimming modalities. This may be due to athletes'

Table 6.2 — Flexindex Male Percentiles (N = 1847)

Age (yr)	N	1	3	5	10	15	17	25	30	35	40	45	50	55	60	65	70	75	83	85	90	95	97	99
5	30	52.3	52.9	53.0	53.9	56.0	56.0	57.0	57.0	58.0	58.6	59.0	60.0	61.0	61.0	61.0	62.0	62.8	63.1	63.7	64.2	66.6	67.0	67.0
6	42	51.4	52.2	53.0	53.1	54.0	54.0	56.0	56.3	57.0	57.0	58.0	58.5	59.0	59.0	60.0	60.0	60.8	64.0	64.0	65.9	69.0	69.0	70.2
7	62	48.0	48.8	49.0	51.1	52.0	52.4	55.0	56.0	56.0	57.0	57.0	57.0	58.0	59.0	60.0	61.0	61.0	64.0	64.0	65.0	66.0	66.0	66.4
8	77	38.5	43.8	47.6	49.0	50.0	50.9	53.0	53.8	54.0	55.0	56.0	56.0	57.0	57.0	58.0	59.0	59.0	62.0	62.0	63.0	63.2	64.7	65.5
9	60	45.0	46.5	47.0	48.0	49.0	50.0	51.0	52.0	52.7	53.0	54.0	54.0	55.0	56.0	57.0	58.3	59.0	59.0	59.2	61.0	62.1	63.0	64.6
10-11	96	39.0	42.6	43.8	45.0	45.0	45.0	47.0	48.0	49.0	49.0	50.0	50.5	51.0	53.0	53.0	53.0	54.0	56.0	57.0	58.0	61.0	62.1	69.0
12-13	73	36.7	37.0	37.6	40.0	41.0	41.0	45.0	46.0	47.0	48.0	48.0	49.0	50.0	50.0	50.8	52.0	53.0	54.0	54.0	55.0	56.4	58.7	63.8
14-15	81	35.0	35.4	36.0	38.0	39.0	39.0	41.0	42.0	43.0	43.0	44.0	45.0	45.0	46.0	47.0	47.0	48.0	51.0	51.0	53.0	53.0	54.6	57.4
16-20	123	31.9	35.7	36.0	38.0	39.0	39.0	40.5	41.0	42.0	42.0	43.0	44.0	45.0	45.2	46.0	47.0	48.0	50.3	51.0	53.0	54.9	55.3	56.8
21-25	84	29.0	30.0	30.2	34.0	35.0	36.0	39.0	40.0	41.1	42.0	42.0	43.5	44.0	44.8	46.0	47.0	48.0	48.9	49.6	52.0	55.9	56.0	58.7
26-30	58	22.3	24.7	30.1	35.0	36.6	37.0	41.0	42.0	43.9	44.0	44.0	46.0	47.0	48.0	49.1	50.9	51.0	54.6	56.0	59.3	62.0	62.0	64.6
31-35	90	22.8	25.3	27.5	31.9	34.4	36.0	37.3	39.0	40.0	40.0	41.1	42.5	43.0	45.0	45.0	46.3	47.8	49.9	51.0	52.1	55.0	56.7	65.2
36-40	111	17.3	22.6	25.5	30.0	31.0	32.0	34.0	35.0	36.0	36.0	38.0	40.0	41.0	43.0	43.5	45.0	47.0	49.3	51.0	53.0	56.0	58.4	61.9
41-45	129	22.3	24.8	25.0	28.0	29.0	30.0	32.0	33.0	34.0	36.0	36.0	37.0	38.0	39.0	40.2	42.0	42.0	43.0	44.0	46.2	49.6	52.6	58.4
46-50	128	20.5	24.8	27.4	29.0	30.0	30.0	31.8	33.0	34.0	35.0	35.0	36.0	37.0	38.2	39.0	41.0	42.0	43.0	44.0	45.0	47.7	48.2	50.7
51-55	152	15.5	17.1	20.6	21.0	23.2	25.0	28.0	30.0	31.0	32.4	34.0	35.0	37.0	38.0	38.2	39.7	41.0	42.3	43.0	44.0	46.0	47.5	49.5
56-60	149	9.0	12.4	16.8	18.0	19.0	19.9	22.0	25.0	29.0	30.0	31.0	32.0	33.0	34.0	35.0	36.0	38.0	41.8	42.0	44.0	47.0	48.0	49.6
61-65	106	14.0	14.5	17.3	14.9	17.0	17.0	20.0	22.0	25.8	27.0	28.0	30.0	31.0	32.0	34.0	35.0	36.0	38.0	38.3	39.0	41.0	41.9	47.0
66-70	80	9.6	11.1	13.0	14.0	16.0	16.5	17.0	18.0	23.0	25.0	25.0	26.0	27.0	28.4	29.4	31.0	32.0	34.1	35.0	36.1	43.0	44.3	45.4
71-75	63	8.6	9.9	12.0	14.0	16.0	15.0	15.0	18.0	19.0	20.0	22.8	24.0	25.0	26.2	27.0	28.4	30.0	31.0	31.7	34.8	36.9	46.3	54.0
76-88	52	9.6	11.0	11.0	12.0	14.8	15.0	15.0	17.0	18.0	19.0	20.0	21.0	21.6	22.0	22.0	23.4	27.0	27.3	29.0	29.8	35.2	37.9	40.9

Table 6.3

Flexindex Female Percentiles (N = 1269)

Age (yr)	N	1	3	5	10	15	17	25	30	35	40	45	50	55	60	65	70	75	83	85	90	95	97	99
5	33	55.3	56.0	56.6	58.0	58.0	58.4	60.0	60.0	61.2	62.0	63.0	63.0	64.0	64.2	65.0	66.4	67.0	67.6	68.0	68.8	71.0	71.2	73.7
6	38	53.0	53.1	53.9	56.7	56.0	57.3	59.0	59.0	60.0	60.0	60.0	61.0	61.0	61.2	62.1	63.0	64.8	65.7	66.9	70.0	70.5	72.7	73.6
7	63	45.5	48.7	51.0	53.0	54.0	54.0	57.0	58.0	59.0	60.0	60.9	61.0	62.0	63.0	64.0	64.0	65.5	68.0	68.0	68.0	71.7	72.1	73.6
8	66	45.0	46.0	48.3	50.5	53.8	54.0	55.3	56.5	57.0	58.0	58.0	59.0	59.8	60.0	60.0	61.0	61.8	63.0	63.0	64.5	66.0	66.2	69.4
9	50	44.5	45.0	45.0	45.9	49.4	50.3	51.3	53.0	54.0	54.0	55.0	55.0	56.0	57.4	58.0	59.0	59.8	61.7	62.0	63.0	64.0	65.1	67.5
10-11	79	39.6	42.0	42.9	46.0	46.0	47.0	48.0	49.0	49.0	50.0	51.0	52.0	52.9	53.0	54.0	54.0	57.0	59.0	60.0	62.0	65.1	66.0	70.2
12-13	69	44.4	47.0	47.0	49.0	50.0	50.0	51.0	51.4	52.0	52.0	52.0	54.0	54.0	54.0	55.0	55.0	56.0	58.0	58.8	61.6	66.0	66.0	69.9
14-15	61	39.8	42.6	43.0	44.0	46.0	46.0	47.0	47.0	48.0	48.0	49.0	49.0	51.0	52.0	53.0	54.0	54.0	56.0	57.0	60.0	66.0	68.2	70.2
16-20	107	37.1	40.2	42.3	43.0	44.0	45.0	46.0	47.0	48.0	49.0	50.0	52.0	52.0	52.6	53.0	53.2	54.0	56.0	56.0	58.0	59.0	60.6	64.8
21-25	78	36.8	39.6	41.9	42.7	44.6	45.0	47.0	47.0	48.0	49.0	50.0	51.0	51.0	52.0	53.0	53.0	54.0	56.0	56.0	57.0	59.0	60.4	64.2
26-30	59	41.6	42.0	42.0	43.0	46.7	47.9	49.0	49.0	50.3	52.0	53.0	53.0	54.9	55.8	57.7	58.6	60.5	64.0	64.0	66.0	67.0	67.3	68.4
31-35	58	34.1	35.0	35.9	38.4	40.0	40.0	43.3	44.0	45.0	46.8	48.0	49.0	50.0	52.0	53.1	54.0	56.5	60.6	62.0	63.3	65.2	66.9	69.4
36-40	84	34.8	36.5	38.2	40.6	43.0	43.0	44.0	46.0	47.0	48.0	48.4	49.0	50.0	50.8	52.0	54.1	56.0	57.9	58.6	59.7	62.0	62.0	67.3
41-45	72	20.1	28.1	29.6	32.4	37.0	37.1	39.8	41.0	43.0	44.0	46.0	47.5	49.0	50.0	51.0	52.0	53.3	55.0	56.4	58.0	61.0	61.9	63.6
46-50	79	19.8	22.7	28.7	32.8	34.0	35.3	39.0	39.4	41.0	43.0	44.0	45.0	45.0	46.0	47.0	48.0	49.0	51.0	51.0	53.0	60.0	60.0	62.7
51-55	79	18.6	20.0	21.9	28.9	32.0	33.0	34.0	36.4	39.0	39.6	40.6	42.5	43.0	44.4	45.0	46.3	47.0	49.6	50.2	52.0	54.0	54.0	56.3
56-60	62	27.2	29.7	30.0	31.0	32.0	32.0	34.3	36.3	38.4	39.4	41.0	42.5	44.0	44.0	45.0	46.0	47.0	49.6	50.9	52.0	56.0	56.2	57.8
61-65	43	19.4	21.6	26.1	28.0	30.6	32.0	34.0	35.0	35.7	36.8	39.7	40.0	41.0	44.0	45.0	45.0	45.5	47.9	48.0	49.8	53.7	54.0	56.9
66-70	38	15.4	16.1	16.9	19.7	30.0	30.0	32.3	34.0	34.0	34.8	36.0	36.5	37.4	38.2	39.1	40.0	42.0	43.7	44.5	45.3	46.2	46.9	53.3
71-75	27	13.3	15.9	17.3	18.6	19.9	20.8	24.5	27.4	30.1	31.4	32.0	33.0	33.6	36.2	37.0	38.0	39.5	41.6	42.1	43.0	44.4	45.4	46.5
76-88	24	11.5	12.4	13.2	14.6	16.0	16.0	17.8	19.8	20.0	20.4	22.0	22.5	23.7	24.0	30.7	32.0	32.0	33.1	33.6	36.8	39.7	42.8	46.9

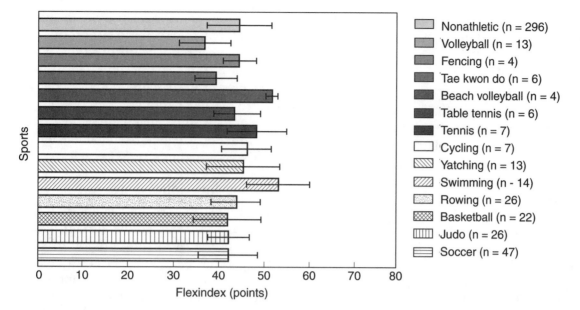

Figure 6.3 Flexindex—sportsmen.

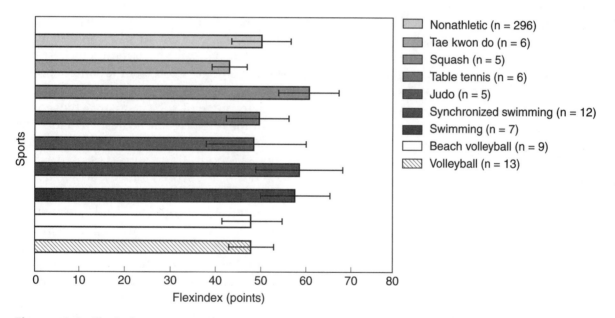

Figure 6.4 Flexindex—sportswomen.

higher muscle mass, which may restrict the amplitude of rotation and posterior shoulder movements. We have found that outstanding sport performance—including Olympic gold medal- and world champion title-winning performances in judo, soccer, and volleyball—may be achieved by athletes with overall Flexindex levels in the range of P_{25} to P_{50} for their age and gender, but with typically higher mobility for some movements specifically relevant for their sport modality. Thus, athletes in some sport modalities clearly need a higher flexibility level than others, such as female synchronized and general swimmers and male swimmers when compared to judoists, yachters, and soccer, volleyball, and basketball players. It should also be noted that athletes practicing the same sport can have significantly different flexibility profiles, depending on their favorite throwing technique (i.e., o-soto-gari versus ippon-seoi-nage in judo) or position (i.e., soccer goalkeeper versus defensive player).

Analyzing Group Data

Flexindex results may also be analyzed with parametric statistics, especially if the sample includes more than 10 or 20 subjects. Use can be made of means and standard deviations or, if a more conservative approach is preferred, either median and range values or interquartile variations.

For comparing Flexindex results from samples, one can choose conventional parametric inferential tests, such as *Student's t-test*, or its nonparametric equivalent, the *Mann-Whitney test*, when the nature of the score distribution is unclear or the sample size is small. When analysis of two samples is needed, use of the *Wilcoxon test* for matched samples (i.e., data sets from testing and retesting) is recommended.

Assessment of the reliability of Flexindex results, similar to the assessment of single movements, can be done by kappa statistics or, preferably, by determining the *intraclass correlation coefficient;* the closer the result is to 1, the higher is the reliability. A valid alternative used to estimate the agreement level among three or more evaluators is *Kendall's correlation coefficient.*

For correlational studies, Flexindex scores may be considered as interval variables and, depending on the statistical categories of other variables, the Pearson moment-product correlation coefficient can be determined.

Movement Analysis

As previously mentioned, there are five possible result groups from a single Flexitest movement, represented by the whole numbers from 0 to 4. The scale was developed to adjust to a Gaussian distribution (in the shape of a bell, with a score of 2 tending to be most frequent, scores of 1 or 3 occasional, and extreme scores

of 0 or 4 rare). In practice, it is somewhat uncommon for all five scores to be observed for a given movement in subjects of a typical small sample of adults. The absolute and relative frequencies of each score, noted as percentages, are the most common descriptions of a movement's results. Results may be grouped and cumulative frequencies obtained, moving from lower to higher scores.

Analyzing Individual Data

Interpreting scores per movement for a given individual is efficiently done through a Flexogram—the graphical presentation of scores for each of the 20 movements as a function of the expected standard for the age and gender as taken from a large population. Each movement's cumulative frequency for the relevant age group and gender, recorded in the data bank, is used to generate the Flexogram, which provides a graphical comparison of results for any Flexitest movement with the distribution in our data bank. Each bar represents results for one movement in 100% of the sample population, with different colors or shades reflecting the percentage of subjects who achieved each one of the five possible scores. The Flexogram in figure 6.5, for example, shows that for movement XV, about 10% of subjects scored a 0, 10% scored a 4, 10% scored a 1, 30% scored a 3, and the remaining 40% scored a 2. For movement XX, about 40% scored a 0 or 1, another 40% scored a 2, and the remaining 20% of subjects scored a 3 or 4.

Sample Flexograms for various male and female major age groups are presented in figures 6.6 to 6.23, as are Flexograms for male soccer, judo, and swimming participants in figures 6.24 to 6.26. From their Flexogram it can be deduced that judoists usually achieved scores of 2 or 3 in Flexitest movements, except for a few subjects who scored 4s in movements II, III, and VII and one athlete who scored a 0 in movement X.

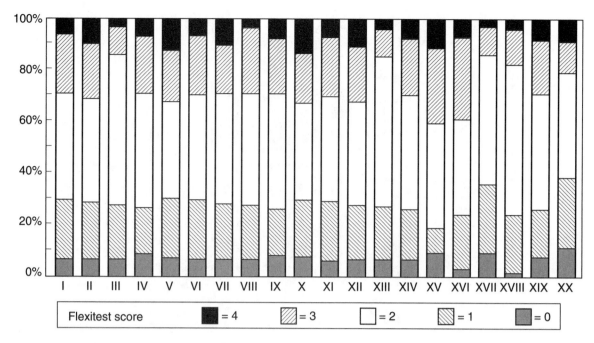

Figure 6.5 Example of a Flexogram.

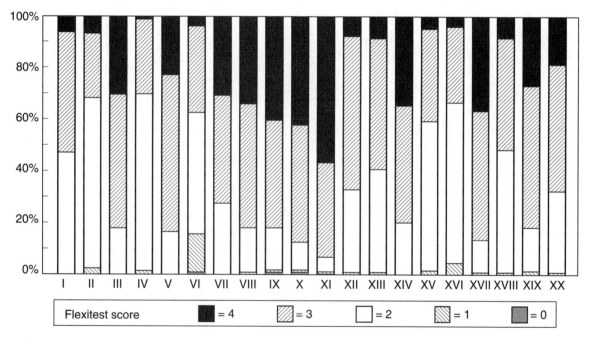

Figure 6.6 Flexogram: 5- to 9-year-old male subjects.

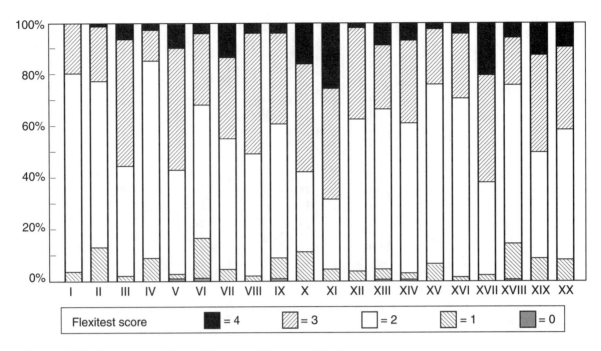

Figure 6.7 Flexogram: 10- to 15-year-old male subjects.

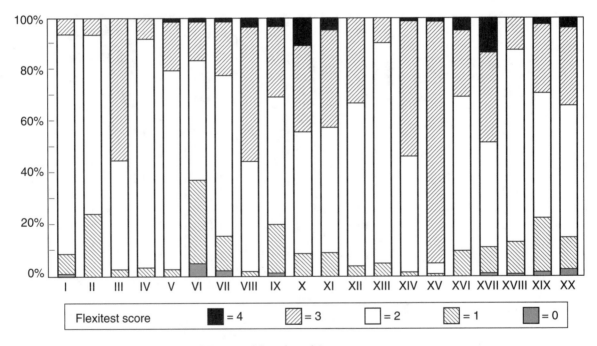

Figure 6.8 Flexogram: 16- to 25-year-old male subjects.

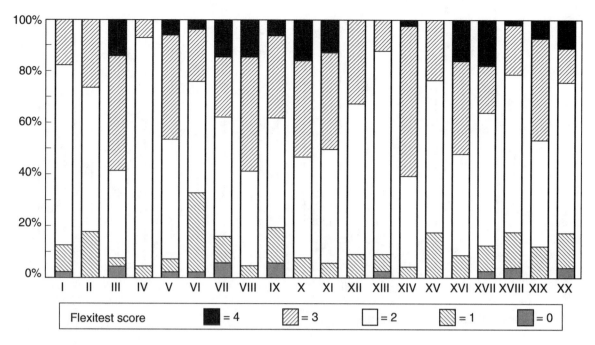

Figure 6.9 Flexogram: 26- to 35-year-old male subjects.

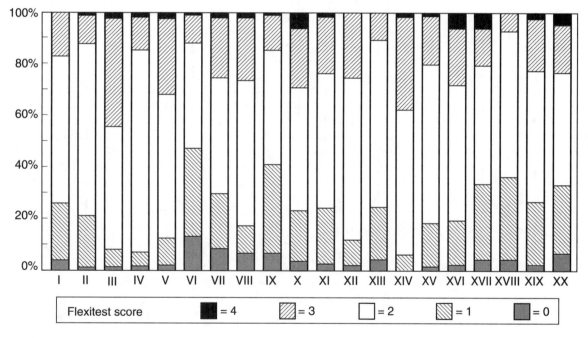

Figure 6.10 Flexogram: 36- to 45-year-old male subjects.

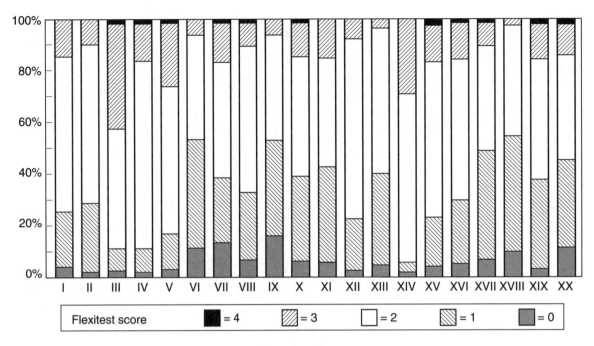

Figure 6.11 Flexogram: 46- to 55-year-old male subjects.

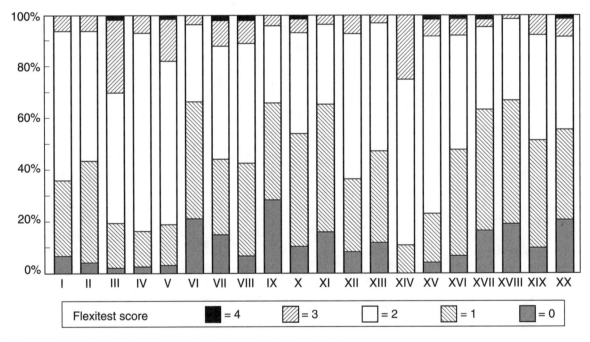

Figure 6.12 Flexogram: 56- to 65-year-old male subjects.

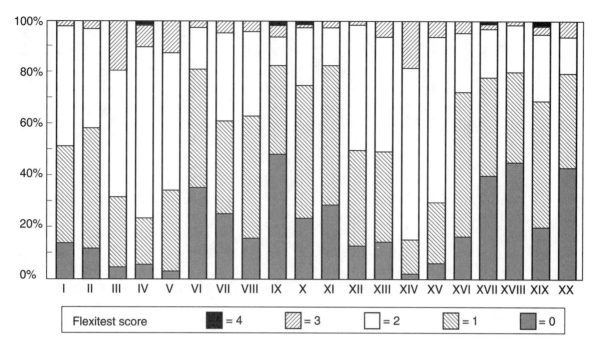

Figure 6.13 Flexogram: 66- to 75-year-old male subjects.

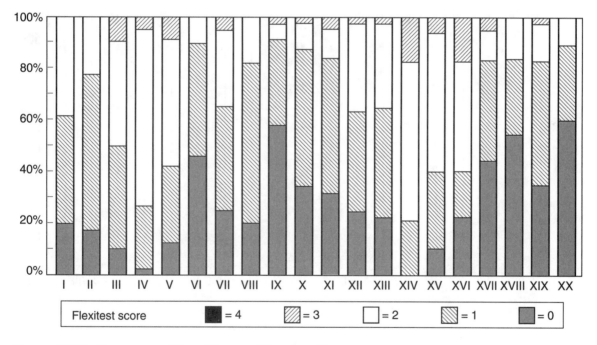

Figure 6.14 Flexogram: 76- to 88-year-old male subjects.

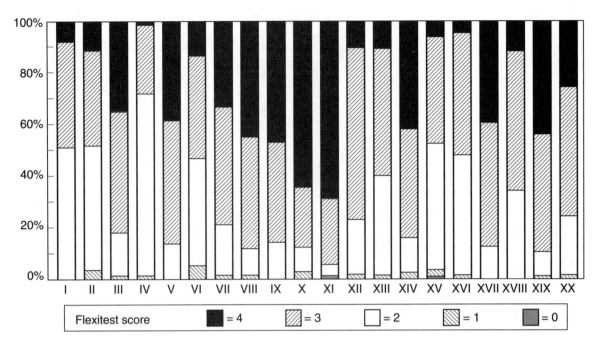

Figure 6.15 Flexogram: 5- to 9-year-old female subjects.

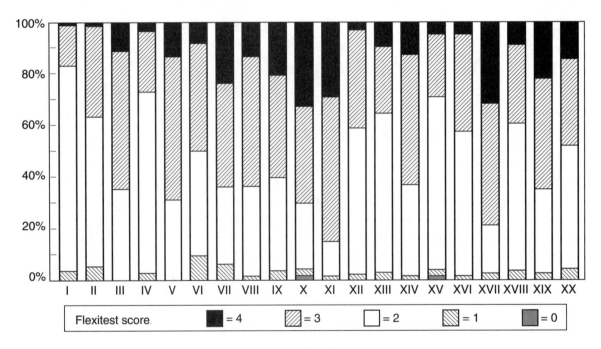

Figure 6.16 Flexogram: 10- to 15-year-old female subjects.

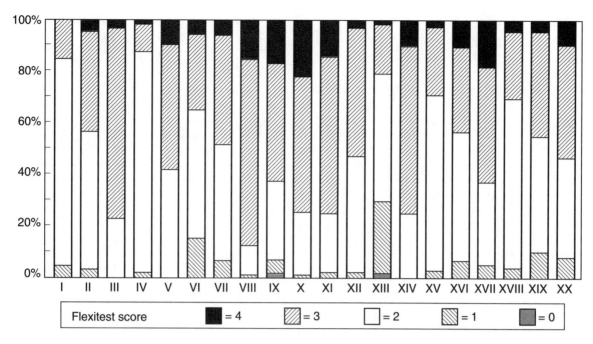

Figure 6.17 Flexogram: 16- to 25-year-old female subjects.

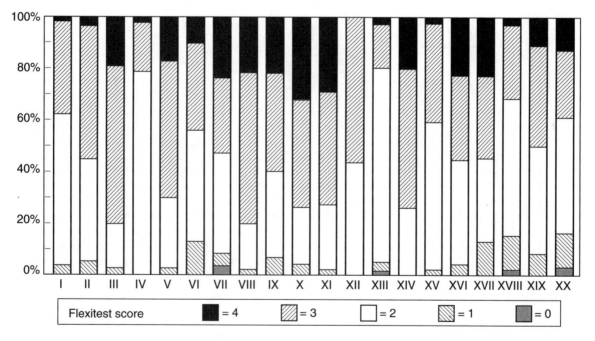

Figure 6.18 Flexogram: 26- to 35-year-old female subjects.

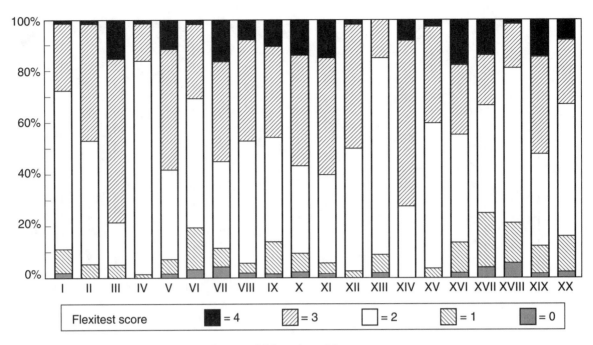

Figure 6.19 Flexogram: 36- to 45-year-old female subjects.

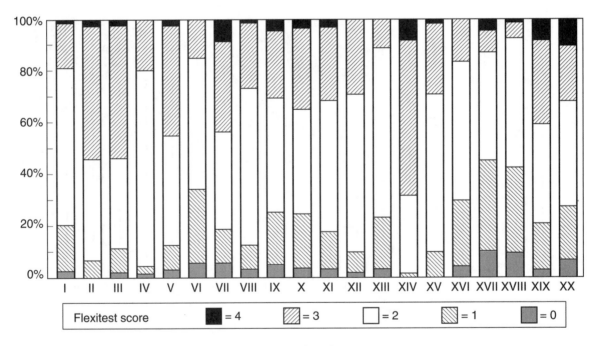

Figure 6.20 Flexogram: 46- to 55-year-old female subjects.

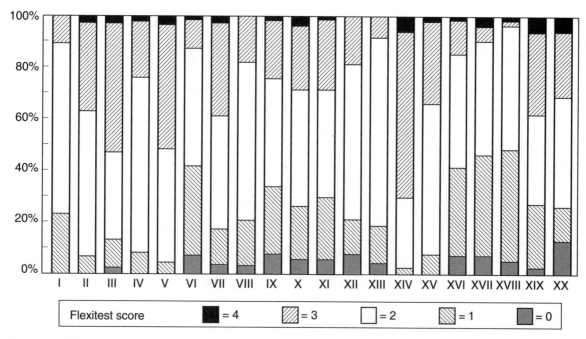

Figure 6.21 Flexogram: 56- to 65-year-old female subjects.

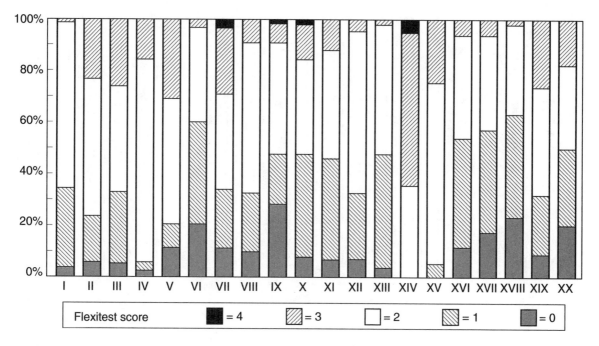

Figure 6.22 Flexogram: 66- to 75-year-old male subjects.

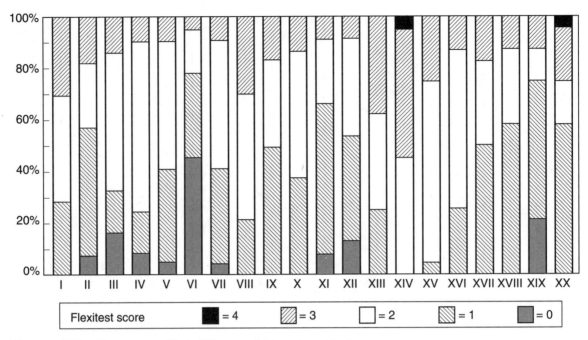

Figure 6.23 Flexogram: 76- to 88-year-old female subjects.

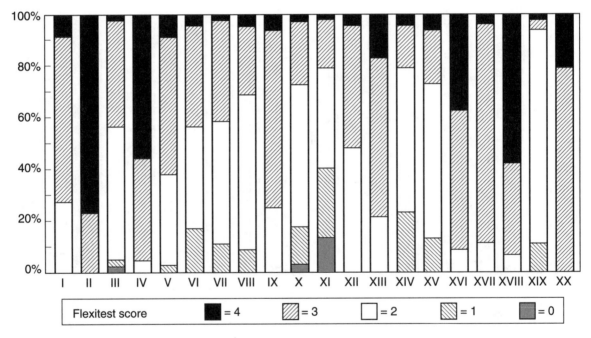

Figure 6.24 Flexogram: male adult soccer players.

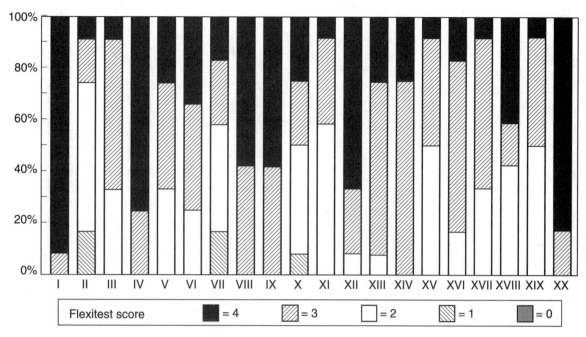

Figure 6.25 Flexogram: male adult judoists.

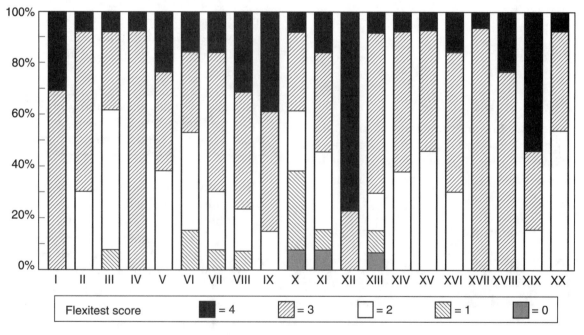

Figure 6.26 Flexogram: male adult swimmers.

Analyzing Group Data

For comparison of movement scores among different groups or for repeated measurements in the same group, chi-square statistics is likely the best and more conservative approach. When working with large samples, the data tend to display a normal distribution and conventional parametric statistical techniques may be used, despite the fact that fractions of unity in means and standard deviation are meaningless.

Comparing Mobility Among Joints

The Flexitest measures the ranges of motion in seven sets of joints (the number in parentheses is the number of movements dedicated to each joint):

1. Ankle (2)
2. Knee (2)
3. Hip (4)
4. Trunk (3)
5. Wrist (2)
6. Elbow (2)
7. Shoulder (5)

Analyzing Individual Data

It is desirable to compare the flexibility of different joints in the same individual; however, as there are differing numbers of movements for the joints, the sums of the scores for individual joints are not directly comparable. A simple mathematical strategy used to permit comparison is to standardize the sums and transform

them into adjusted scores by multiplying each sum by 20 and then dividing the result by the number of movements for that specific joint. For instance, for the ankle there are two movements; thus, the two scores are added, the sum is multiplied by 20 (total number of movements), and the total is divided by two (the number of movements for that joint). This final result is directly comparable to the overall Flexindex score and the other joints' adjusted scores. Table 6.4 shows two examples to illustrate how flexibility can vary in two people who have the same Flexindex score. The standardized joint scores reveal the specific differences.

Subjects A and B have the same overall flexibility level; however, their mobility profiles per joint are significantly different. Subject A's knee and hip flexibility are slightly higher than his overall flexibility, and his shoulder mobility exhibits a relatively substantial reduction. Subject B shows reduced mobility in the lower limbs and a relatively higher mobility in the upper limbs. Alternatively, you may further simplify analysis to five joint sets by grouping the ankle and knee results and the wrist and elbow data.

Individual Flexindex scores (the sum of all 20 movement scores) should always be analyzed against age- and gender-reference norms.

The standardized weighted score for each joint permits the evaluator to immediately compare mobility in the different joints of one person. Higher scores reflect higher mobility and lower scores, lower mobility. If significant discrepancies arise among the standardized weighted scores for the different joints, concomitant analysis of the variability indexes, which will be discussed later, may add useful information for clinical or sport performance purposes.

Analyzing Group Data

For comparison of flexibility scores in specific joints among different groups or for repeated measurements in the same group, the evalu-

Table 6.4	Example of Joint Score Disparity With Identical Flexindex	
Source	Subject A	Subject B
Flexindex	40	40
Standardized joint score		
Ankle	40	30
Knee	50	30
Hip	45	25
Trunk	47	40
Wrist	40	50
Elbow	40	50
Shoulder	28	52

ator can use parametric inferential techniques, including t-tests and analyses of variance, since these data tend to have a normal distribution.

Analyzing the Flexibility Homogeneity Profile

According to the theory of specificity of flexibility, it is very rare for a person to achieve identical scores in all of the Flexitest movements. Standardized joint scores were introduced to allow for the direct comparison of mobility among the seven joints. This approach is not sufficient to disclose all the different possibilities for heterogeneity in Flexitest measurements in a given individual, however. In theory, the same Flexindex result may originate from different combinations of results for each movement. For instance, a Flexindex score of 40 may be conferred if all movements were rated 2s or if 10 movements were rated 1s and the other 10 were rated 3s. While the Flexindex is the same in both instances, the *passive joint mobility profile* for each is quite different. Therefore, assessment of overall flexibility should consider not only the Flexindex, but also the profile of individual movements' ranges of motion.

The profile of flexibility-measurement homogeneity for a subject may be assessed using five dimensionless indexes, all of them belonging to a ratio scale of measurement specifically calculated for this purpose:

1. Intermovement variability index (IMVI)
2. Interjoint variability index (IJVI)
3. Flexion–extension variability index (FEVI)
4. Between-segments variability index (BSVI)
5. Distal–proximal variability index (DPVI)

The homogeneity of joint mobility scores rated by these specific indexes—IMVI, IJVI, FEVI, BSVI, and DPVI—is highly important for interpreting individual Flexitest results. The higher the measurement's heterogeneity, the more likely you are to find that significant joint mobility limitations exist. Some variability scores are substantially different between athletes and nonathletes, reflecting specific profiles

of body flexibility that may correspond to the characteristics of the sports practiced. Although these indexes are dimensionless and measured in a ratio scale, their distribution is somewhat asymmetrical, imposing some limitations on the use of parametric statistics for comparing index results in two or more samples or measurements. In light of that, we prefer to use percentile ranges rather than standard deviations of the means in interpreting these variability indexes.

Interestingly, the first two indexes—IMVI and IJVI, which evaluate the dispersion of measurements around the mean of scores—are not influenced by age, gender, or the Flexindex score, which makes them very easy to interpret. Since the other three indexes are calculated ratios, they tend to present values around unity. By attributing expressions to these results, people can be categorized as having a homogeneous or normal, a somewhat atypical, or a highly atypical flexibility profile (the latter two expressions having categories for both more and less flexible than normal) as a function of each one of the variability index results (see table 6.5).

Intermovement Variability Index (IMVI)

The IMVI represents the *variability of the scores of different movements, regardless of the joint's or the movement's kinesiological type*. It is the standard deviation of the individual scores for the 20 movements. Theoretically, IMVI can range from 0 to 2, but seldom is it higher than 1. Mean IMVI is approximately 0.65 for both males and females, and in two-thirds of individuals it ranges from 0.40 to 0.90. A child with 5 scores of 1, 5 scores of 3, and 10 scores of 2 for Flexitest's 20 movements has a Flexindex of 40 and an IMVI of 0.63. High IMVI values are normally related to the presence of abnormally low scores in one or two movements, particularly in children or athletes, often due to the effects of a previous or current motion-restraining injury.

Interjoint Variability Index (IJVI)

The IJVI reflects the degree of *variability in the range of motion among the different joints assessed* with the Flexitest. IJVI is established by calculating the standard deviation of the mean

Table 6.5		Reference Values Based on Percentile Ranges For Flexitest's Variability Indexes				
Variability	Percentile	IMVI	IJVI	FEVI	BSVI	DPVI
Male						
Very atypical	$< P_5$	< 0.44	< 0.21	< 0.80	< 0.74	< 0.74
Atypical	P_5-P_{17}	0.44–0.54	0.21–0.28	0.80–0.93	0.74–0.86	0.74–0.87
Normal	$P_{18}-P_{83}$	0.55–0.78	0.29–0.54	0.94–1.29	0.87–1.25	0.88–1.30
Atypical	$P_{84}-P_{95}$	0.79–0.85	0.55–0.64	1.30–1.71	1.26–1.64	1.31–2.09
Very atypical	$> P_{95}$	> 0.85	> 0.64	> 1.71	> 1.64	> 2.09
Female						
Very atypical	$< P_5$	< 0.48	< 0.21	< 0.83	< 0.78	< 0.75
Atypical	P_5-P_{17}	0.48–0.54	0.21–0.28	0.83–1.00	0.78–0.88	0.75–0.83
Normal	$P_{18}-P_{83}$	0.55–0.79	0.29–0.54	1.01–1.29	0.89–1.18	0.84–1.13
Atypical	$P_{84}-P_{95}$	0.80–0.88	0.55–0.64	1.30–1.51	1.19–1.44	1.14–1.74
Very atypical	$> P_{95}$	> 0.88	> 0.65	> 1.51	> 1.44	> 1.74

IMVI: Intermovement Variability Index; IJVI: Interjoint Variability Index; FEVI: Flexion-Extension Variability Index; BSVI: Between-Segments Variability Index; DPVI: Distal-Proximal Variability Index.

scores for the movements of each of a subject's seven joint sets. The IJVI mean value is close to 0.4 for males and females; 70% of results range from 0.25 and 0.55. A student whose ankle, hip, trunk, wrist, and shoulder movements scored 2s, elbow movements scored 1s, and knee movements scored 3s, thus achieving a Flexindex of 40 points, will have an IJVI score of 0.53. High IJVI values are typically found in subjects who have an injured, limited-mobility, or hyperlax joint, while low values are more often seen in untrained subjects with about-average levels of joint mobility.

Flexion–Extension Variability Index (FEVI)

The FEVI *compares mobility in flexion and extension movements.* FEVI is calculated as a ratio between the means of scores obtained for all flexion and extension movements in the ankle, knee, hip, trunk, wrist, and elbow. If the Flexitest scoring system shows there is similar mobility in flexion and extension movements, the FEVI value is about 1. Even though results theoretically can have positive values over a wide scale, in practice almost all values range

from 0.5 to 2. There is a very small and insignificant difference between males and females for this index. When only the FEVI is high, there is a predominance of flexion over extension mobility, which usually is associated with a loss of muscle tonus and important joint stiffness, as are often seen in sarcopenic individuals. This standard is still ordinarily seen in small children, especially those of less than one year of age, due to the predominance of flexor over extensor muscles. In some pathological cases, such as in patients with spasticity, significant discrepancies in flexion and extension mobility may occur and be identified by FEVI scores that are far from unity.

Between-Segments Variability Index (BSVI)

The BSVI compares *passive mobility between the lower and upper body.* BSVI is calculated as a ratio between the means of eight lower-limb and nine upper-limb movements. If mobility is equal in both the upper and lower limbs, the result is 1. When Flexitest-assessed passive mobility is higher for the lower limbs, the BSVI is higher than 1; the opposite applies for greater upper-limb mobility. Values may vary considerably, but in most cases they are above unity. If all scores from a segment were zero, which is extremely rare, the BSVI value would be infinite and could not be determined. The trend for BSVI values in our lab's data are average scores of 1, with slight differences for gender that increase with age. Mean values of 1.07 and 1.04 were found for males and females respectively, with scores for two-thirds of males and females ranging from 0.86 to 1.25 and from 0.88 to 1.18 respectively. When the BSVI exceeds 1, it is due to a predominance of passive mobility amplitude in one of the segments, whether in the lower limbs (when the BSVI is much higher than 1) or the upper limbs (when the BSVI is much lower than 1). A high BSVI value could be a sign of muscular dimorphism because it is sometimes found in people who are much more developed and stronger (and often less flexible as well) in the upper body than in the lower limbs.

Distal–Proximal Variability Index (DPVI)

The DPVI characterizes *differences between the maximum passive mobility of limbs' distal and proximal joints*. DPVI is calculated as the ratio of the mean scores of eight distal movements in the ankle, knee, wrist, and elbow and the mean scores of nine proximal movements in the hip and shoulder. Like the previous index, if a subject's flexibility is homogeneous, the DPVI is equal to 1. When the distal passive mobility exceeds the proximal, the DPVI goes higher, and vice versa. The values are always positive, starting at 0.01 and eventually reaching quite high values such as 10 or even higher. When the proximal score is zero and it is impossible to calculate the index or when its magnitude exceeds 10, for further statistical analysis the result is considered to be 10. DPVI is directly related to age ($r > 0.36$ and < 0.46) and inversely related to Flexindex ($r < -0.48$ and > -0.60). Among children and adolescents (up to 15 years of age), values higher than the mean for each gender are quite rare (5 to 10% of the cases). Most subjects over age 60 have DPVI results higher than the mean for each gender, 1.15 for males and 1.04 for females. When comparing two individuals with similar Flexindex results, the subject with the higher DPVI will tend to be older. High DPVIs are often seen in old, sedentary people and tend to decrease after stretching training. On the other hand, an infant-juvenile mobility pattern exists, reflecting lower distal compared to proximal passive mobility and consequently conferring a low DPVI. Finally, a high DPVI shows that the passive mobility of distal joint movements—in the ankle, knee, wrist, and elbow—quite exceeds the passive mobility of proximal joints, i.e., the hip and shoulder.

Athletes have a more heterogeneous flexibility profile per movement than nonathletes do, a fact that is clear from all five variability indexes. The one exception is DPVI in females, in which athletes and nonathletes show similar results. Thus, higher flexibility heterogeneity profiles are often found in those devoted to sports competition. In such subjects, the higher flexibility levels were found in the joints performing biomechanically relevant movements for the sports' performance, which varied according to the sports modality, characteristics, and playing positions.

Final Statistical Considerations

In this chapter, different approaches were proposed and examples were given for statistically based analyses of Flexitest data. Flexindex, individual movement, and homogeneity profiles were covered in detail and age and gender reference norms were provided for comparison. An example is presented below to illustrate the application of some of the techniques discussed. Additional examples are presented as case studies in chapter 9.

Table 6.6 presents Flexitest data for a 36-year-old woman, who is moderately active. With this data, the appropriate percentile curve or table, and the Flexogram for 36- to 55-year-old females, it is possible to detect that the subject's overall flexibility is somewhat low, between the 25th and 40th percentiles (much closer to 25 than to 40), and should be improved as a whole. Analysis of each movement allows the identification of specific limitations that should be initially tackled.

Table 6.6		Flexitest Data for a 36-Year-Old Moderately Active Woman	
Movement	**Score**	**Movement**	**Score**
I	0	XI	2
II	2	XII	2
III	2	XIII	1
IV	3	XIV	2
V	3	XV	2
VI	2	XVI	3
VII	3	XVII	3
VIII	3	XVIII	3
IX	2	XIX	3
X	1	XX	2
Flexindex		44	

The subject's dorsiflexion is extremely limited, and her hip, trunk, and wrist extension are somewhat lower than is typical for women in this age group. On the other hand, there is above-average mobility for shoulder posterior adduction and for hip flexion, which may facilitate some specific sports movements. The most common cause for comparably higher restriction in one of the ankle movements in women is the routine use of high-heeled shoes, which may lead to chronic shortening of ankle muscles and limited mobility. The relative shortness in extension mobility in other movements is probably due to the lack of specific training and to little use being made of related muscle groups; lessened wrist extension is an ordinary finding in tennis players who must firmly grip their rackets to execute backhand and forehand shots. Finally, increased mobility of these otherwise limited movements may allow better serving and reaching for short balls close to the net. This flexibility data profile is compatible with a long history of tennis practice and routine use of high heels by a 36-year-old woman.

In figure 6.27, we present a logical and simple sequencing of the recommended steps in performing Flexitest data analysis, with the aim of guiding the reader in using the material presented in this chapter.

When interpreting Flexitest results:

1. Start interpreting the results with Flexindex. Analyze its absolute value to see whether it points to hypo-, hyper-, or normal mobility. Next, compare the value to the age and gender reference norms to determine the Flexindex corresponding percentile.

2. Consult variability indexes to identify values that deviate from normal ranges.

3. Next, make a comparative analysis for joints and, finally, for each movement, based on absolute results and on the proportion of that score for age group and gender.

4. Ideally, one should provide the subject with objective feedback when the measurements are completed, and this is best done by referencing the Flexindex score. Do not fear presenting the result as a percentile for age and gender; even children, to the wonder of some, understand the meaning of being ranked in relation to 100 colleagues of the same age and gender.

5. Try to indicate in your report which movements the subject should work on to improve their maximum ranges of motion. Focus not only on improving to reach reference values for age and gender, but also improving those movements that suit the specific purposes of the subject.

6. Bear in mind that hypermobility typically is not a goal to be achieved, and may even be deleterious for performance in some sports (for instance, ankle plantar hyperflexion in adult runners) or some daily activities.

Figure 6.27 Interpreting Flexitest results.

PART III

RESEARCH ON AND APPLICATIONS FOR FLEXITEST

Flexitest Research

7

Since its development, the Flexitest has been used in many research situations by the author and other investigators. As detailed in the previous chapter (pages 111-136), a considerable amount of data has been collected during recent years, and part of this material remains unpublished. Table 7.1 lists most of the Flexitest scientific materials presented and published in different formats and languages. This table also includes the Web address from which a paper's full text or abstract, if available, can be freely downloaded. The languages in which each paper's abstract and full text were published are also noted in table 7.1.

This chapter presents some of the research that has made use of the Flexitest, focusing primarily on the studies in which our research group has directly participated in collecting and interpreting the results. For easy reference while reading, the number of the corresponding item in table 7.1 is cited in the text (e.g., [4]). The studies are described not in chronological order, but in groups according to the nature or characteristics of the study. The studies are classified in five different categories:

1. Reliability studies
2. Concurrent validity studies
3. Strictly methodological studies
4. Observational studies
5. Interventional studies

For each study, we introduce the research question and objectives. Next, brief descriptions of the sample, methodological strategy, and main results of the study are presented. Last, a concise interpretation is made of the findings and their implications for flexibility research and intervention. We believe that reading the summary of the findings will help the reader gain a better understanding of the scientific basis and evolution of the Flexitest over the last 20 years, possible applications for the test, interesting relationships between flexibility and other variables, and the body's response to interventions such as regular physical activity and warm-up.

Reliability Studies

In the field of measurement and evaluation, reliability issues are of utmost importance. Reliability is a concept involving multiple aspects that must be considered when using the Flexitest for flexibility assessment. Ideally, consecutive measurements obtained in the absence of intervention should provide identical values. However, potential sources of error are always present and should be precisely identified and quantified.

Interobserver Reliability

Data originally from [14] and partially reanalyzed for this summary.

▾ **Rationale:**

Reliability is related to the stability of the results of a test (i.e., the degree of consistency found in the results). It is typically expressed as a correlation coefficient between two or more series of independent results from a particular group of subjects, and the higher the coefficient, the higher the reliability of a test. Ideally, the independently assessed scores should be identical, but that does not actually happen in practice. There will always be discrepancies, and it is up to the researcher to identify their sources. Depending on the study's design, one may test reliability among two or more evaluators' scores or among two

Table 7.1

	Author(s)	Type	Source	Details
1	Araújo CGS, Pável RC, Pina-Almeida A	Presented at Regional Brazilian Congress of Sport Sciences	Congress Proceedings	1980
2	Pável RC, Araújo CGS	Presented at Regional Brazilian Congress of Sport Sciences	Congress Proceedings	1980
3	Araújo CGS, Perez AJ, Pável RC	Presented at AIESEP World Congress	Rev Artus	1981
4	Pável RC, Araújo CGS	Presented at AIESEP World Congress	Rev Artus	1981
5	Araújo CGS, Perez AJ, Haddad PCS	Presented at 2nd Brazilian Congress of Sports Sciences	Rev Bras Ciên Esporte	1981
6	Haddad PCS, Perez AJ, Araújo CGS	Presented at 2nd Brazilian Congress of Sports Sciences	Rev Bras Ciên Esporte	1981
7	Araújo CGS, Perez AJ, Haddad PCS, Pável RC	Presented at Regional Brazilian Congress of Sport Sciences	Congress Proceedings	1982
8	Perez AJ, Araújo CGS	Presented at Regional Brazilian Congress of Sport Sciences	Congress Proceedings	1982
9	Araújo CGS	Full paper	Medicina do Esporte	1983; 7(3-4):7-24.
10	Perez AJ, Araújo CGS	Presented at 36th Congress of Brazilian Society for Advancement of Sciences	Ciência e Cultura	1984
11	Araújo CGS, Haddad PCS	Full paper	Comunidade Esportiva	1985; 35:12-17
12	Araújo CGS, Perez AJ	Full paper	Boletim da Federação Internacional de Educação Física	1985; 55(2):20-31
13	Araújo CGS	Full paper	Kinesis	1986; 2:251-267

Flexitest: List of Scientific Materials		
Title	**Language**	**Internet**
Flexiteste – análise preliminar de sua objetividade e confiabilidade	Abstract in Portuguese	www.clinimex.com.br/abstracts
Flexiteste – nova proposição para avaliação da flexibilidade	Abstract in Portuguese	www.clinimex.com.br/abstracts
Correlação entre flexibilidade segmentar e geral em alunos de Educação Física	Abstract in Portuguese	www.clinimex.com.br/abstracts
Flexiteste – método de avaliação da amplitude máxima de 20 movimentos articulares	Abstract in Portuguese	www.clinimex.com.br/abstracts
Comparação entre a amplitude máxima de flexão e de extensão em seis articulações	Abstract in Portuguese	www.clinimex.com.br/abstracts
Comparação da flexibilidade dos dimídios corporais em alunos de Educação Física	Abstract in Portuguese	www.clinimex.com.br/abstracts
Resultados práticos de um curso teórico-prático de 18 horas em flexibilidade na aplicação do Flexiteste	Abstract in Portuguese	www.clinimex.com.br/abstracts
Características cineantropométricas de deficientes mentais	Abstract in Portuguese	www.clinimex.com.br/abstracts
Existe correlação entre flexibilidade e somatotipo? - uma nova metodologia para um problema antigo	Text in Portuguese	NA
Características de flexibilidade em escolares e pré-escolares dos dois sexos	Abstract in Portuguese	www.clinimex.com.br/abstracts
Efeitos do aquecimento ativo sobre a flexibilidade passiva	Text in Portuguese	NA
Características da flexibilidade em pré-escolares e escolares dos dois sexos	Text in Portuguese	NA
Flexiteste: uma nova versão dos mapas de avaliação	Text in Portuguese	NA

(continued)

Table 7.1

	Author(s)	Type	Source	Details
14	Araújo CGS	Presented at 8th Brazilian Congress of Sports Medicine	Congress Proceedings	1987
15	Araújo CGS	Ph.D. dissertation	Univ Federal do Rio de Janeiro	1987, 440 p.
16	Araújo CGS, Nóbrega ACL	Presented at 9th Brazilian Congress of Sports Medicine	Congress Proceedings	1989
17	Oliveira Jr AV, Araújo CGS	Presented at 45th Brazilian Congress of Cardiology	Arq Bras Cardiol	1989
18	Araújo CGS	Presented at 10th Brazilian Congress of Sports Medicine	Congress proceedings	1991
19	Araújo CGS	Presented at 10th Brazilian Congress of Sports Medicine	Congress proceedings	1991
20	Farinatti PTV	M.Sc. thesis	Universidade Federal do Rio de Janeiro	1991, 132 p.
21	Farinatti PTV, Soares PPS, Vanfraechem JHP	Full paper	Sport	1995; 4:36-45
22	Farinatti PTV, Araújo CGS, Vanfraechem JHP	Full paper	Science et Motricité	1997; 31:16-20
23	Carvalho ACG, Paula KC, Azevedo TMC, Nóbrega ACL	Full paper	Rev Bras Med Esporte	1998; 4(1):2-8
24	Farinatti PTV, Nóbrega ACL, Araújo CGS	Full paper	Horizonte (Lisboa)	1998; 14(82):23-31
25	Araújo CGS	Presented at 1999 ACSM Annual Meeting	Med Sci Sports Exerc	1999; 31(5-Suppl):S115
26	Araújo CGS	Book chapter	O Exercício	São Paulo: Atheneu, 1999, p. 25-34.
27	Araújo CGS	Full paper	Rev Bras Ciên Mov	2000; 8(2):25-32

Title	Language	Internet
Flexiteste – fidedignidade intra e inter-observadores	Abstract in Portuguese	www.clinimex.com.br/abstracts
Medida e avaliação da flexibilidade: da teoria à prática	Text in Portuguese and abstract in English	www.clinimex.com.br/texts
Características da flexibilidade de atletas brasileiros de elite	Abstract in Portuguese	www.clinimex.com.br/abstracts
Melhora da flexibilidade de pacientes submetidos a programas de reabilitação cardíaca	Abstract in Portuguese	www.clinimex.com.br/abstracts
Comportamento circadiano da flexibilidade: uma avaliação pelo Flexiteste	Abstract in Portuguese	www.clinimex.com.br/abstracts
Comparação de três métodos de avaliação da mobilidade articular em indivíduos sadios	Abstract in Portuguese	www.clinimex.com.br/abstracts
Estudo da aplicabilidade do trabalho de flexibilidade em Educação Física: uma abordagem multidisciplinar	Text in Portuguese and abstract in English	www.clinimex.com.br/texts
Influence de deux móis d'activités physiques sur la souplesse de femmes de 61 à 83 ans à partir d'un programme de promotion de la santé	Text in French and abstract in English	NA
Influence of passive flexibility on the ease for swimming learning in pre-pubescent and pubescent children	Text in English and abstract in French	NA
Relação entre flexibilidade e força muscular em adultos jovens de ambos os sexos	Text in Portuguese and abstract in English	NA
Perfil da flexibilidade em crianças de 5 a 15 anos de idade	Text in Portuguese	NA
Body flexibility profile and clustering among male and female elite athletes	Abstract in English	NA
Avaliação e treinamento da flexibilidade	Text in Portuguese	NA
Correlação entre diferentes métodos lineares e adimensionais de avaliação da mobilidade articular	Text in Portuguese and abstract in English	www.clinimex.com.br/texts

(continued)

Table 7.1				
	Author(s)	**Type**	**Source**	**Details**
28	Coelho CW, Araújo CGS	Full paper	Revista Brasileira de Cineantropometria & Desempenho Humano	2000; 2(1):31-41
29	Silva LPS, Palma A, Araújo CGS	Full paper	Rev Bras Ciên Mov	2000; 8(3):15-20
30	Araújo CGS	Full paper	Sports & Medicine Today	2001; 1(2):34-37
31	Chaves CPG	M.Sc. thesis	Universidade Gama Filho	2002, 140 p.
32	Araújo CGS	Full paper	Rev Bras Med Esporte	2002; 8(1):13-9
33	Araújo DSMS, Araújo CGS	Full paper	Rev Bras Med Esporte	2002; 8(2):37-49
34	Chaves CPG, Simão Jr RF, Araújo CGS	Full paper	Rev Bras Med Esporte	2002; 8(6):212-218
35	Araújo CGS, Oliveira AJ, Almeida MB	Presented at 2002 Sports Sciences Symposium— São Paulo, Brazil	Rev Bras Ciên Mov	2002; 10 (Supl):

or more measurements taken by the same evaluator.

▼ Objective:

The objective of this study was to establish a preliminary measure for interobserver reliability when measuring flexibility with the Flexitest.

▼ Sample:

Twenty-nine undergraduate physical education students (19 men and 10 women) volunteered to take part in the study. All were familiar with the Flexitest prior to the study. They were moderately active, but did not take part in competitive sports. None had locomotor impairments that made measuring the maximum ranges of motion of the different movements difficult or impossible.

▼ Methods:

All of the subjects were administered to the Flexitest on the same day during one session. A subject was assessed by one evaluator and then immediately afterward by the other, and the evaluators alternated between being the first to administer the test. The evaluators wrote the scores, without telling them to the subjects. The study was designed so that the evaluators did their own measuring and scoring without knowing the other's results.

▼ Results:

The results distribution and Flexindex central tendency were similar for both evaluators ($p < 0.05$). There was a high association of individual values, with an intraclass correlation coefficient of 0.89 ($p < 0.01$) (figure 7.1).

		(Continued)
Title	**Language**	**Internet**
Relação entre aumento da flexibilidade e facilitações na execução de ações cotidianas em adultos participantes de programa de exercício supervisionado	Text in Portuguese and abstract in English	NA
Validade da percepção subjetiva na avaliação da flexibilidade de adultos	Text in Portuguese and abstract in English	www.clinimex.com.br/texts
Flexitest – an office method for evaluation of flexibility	Text in English	www.clinimex.com.br/texts
Variabilidade da flexibilidade em mulheres adultas: exemplos de abordagens transversal e longitudinal	Text in Portuguese and abstract in English	www.clinimex.com.br/texts
Flexiteste: proposição de cinco índices de variabilidade da mobilidade articular	Text in Portuguese and abstract in English	www.clinimex.com.br/texts
Autopercepção das variáveis da aptidão física	Text in Portuguese and abstract in English	www.clinimex.com.br/texts
Ausência de variação da flexibilidade durante o ciclo menstrual em universitárias	Text in Portuguese and abstract in English	www.clinimex.com.br/texts
É apropriado utilizar versões condensadas do Flexiteste?	Abstract in Portuguese	www.clinimex.com.br/abstracts

When assessing the results of the individual movements, the evaluators disagreed on the scores for 205 (35%) of the 580 measurements (29 subjects × 20 movements); the difference was greater than one point in just under 10% of these. However, each evaluator gave a lower score than the other did about the same number of times (103 versus 102; $p > 0.05$), so neither evaluator systematically underestimated flexibility in any of the movements.

The mean proportion of concordance was 65%. That number was even higher for the distal joints. The largest number of disagreements occurred for shoulder medial rotation in movement XX.

▼ **Conclusion:**

There is some contradiction between the excellent intraclass correlation coefficient ob-tained with Flexindex and the similarity of distributions on the one hand and, on the other, the relatively low proportion of agreement in individual movement scores found in the study.

Careful review of the results revealed a clear tendency for the same movement to be scored higher in the second measurement than in the first. Therefore, it seemed that the second measurement was "contaminated" by the first due to a "warm-up" effect on the subject; this effect is also seen in other flexibility measurement tests (Atha and Wheatley 1976; Ellis and Stowe 1982). This hypothesis is also support-ed by the fact that higher discrepancies were seen in movements where muscular mass may restrain the range of motion, such as in the hip, trunk, and shoulder. Another factor that

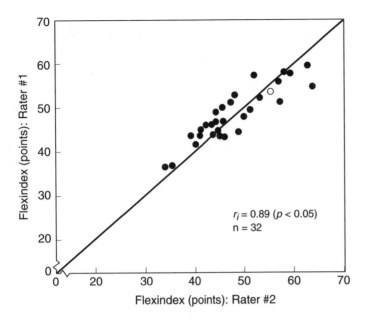

Figure 7.1 Flexindex reliability between two expert evaluators. The unfilled circle represents two points that overlap.

may have improved mobility in the second measurement is the higher stretch tolerance of students (Magnusson et al. 1996). The most important difference was seen in movement XX, for which the two evaluators had differently interpreted the proposed methodology. This pilot study pointed out some problems that existed in determining Flexitest reliability.

Intraphoto Reliability

Data originally from [1, 13-15] and partially reanalyzed for this summary.

▼ Rationale:

As shown in the pilot study, one of the potential limitations in determining the intra- and interobserver reliability of a test or procedure is the biological variability of the subjects under assessment. This variable can be controlled by replacing the subjects with photographs of subjects in the positions to be evaluated.

▼ Objective:

The objective of this study was to determine and discuss intra- and interobserver reliability in Flexitest assessment when the variability inherent in subjects' movements is controlled by using photographs.

▼ Sample:

Two series of photographs were taken of subjects as the Flexitest was administered. The first series consisted of 788 photographs, 35 to 40 of each of the Flexitest's 20 movements performed by very young female dancers. The second series had 260 photographs of 13 complete Flexitests performed by undergraduate students in physical education and by subjects who were regularly active at a health fitness center. The flexibility level displayed an intentionally high variability.

▼ Methods:

The first series of photographs encompassed nearly all of the possible scores, but slightly favored extreme results. The purpose of the second series was to reflect data gathered in more conventional situations. For the first series of photos, all possible scores for each movement were obtained, without regard for logical sequence. Next, all the photos of a given movement were printed in black and white on a single sheet of photographic paper measuring about 8-by-11 in; the photos, each measuring 24-by-35 mm, were numbered sequentially for the assessment. The same process took place for the second series of photos, which presented in sequence all the Flexitest movements of a subject so it was possible to evaluate the complete Flexitest of the first subject before going on to that of the second. Again, the individual movements' scores were written down before the results were considered as a whole.

To establish intraobserver reliability, scoring of each Flexitest movement photo in the two series was performed twice, a week apart. After analyzing the spreadsheet results and calculating the appropriate indexes, the photos that received different scores in the two assessments were reanalyzed to establish a final score; this final score was then used in the analysis of interobserver reliability. Reliability between observers was determined by comparing the measurements of two evaluators, one who was quite experienced in using the technique and the other not, in order to check the influence of experience on intraobserver reliability. Before the actual assessment, the evaluators reviewed the Flexitest methodology and scored a small number of extra photos that did not belong to either of the series. All photos were taken by a professional photographer, using a Nikon Nikkormat FTn camera with a 50mm lens set at a 1.4 aperture and using ASA 400 black-and-white film to prevent angle distortion and permit natural light to be used.

▾ **Results:**

The intraobserver reliability study for the first series of photographs demonstrated intraclass correlation coefficients ranging from 0.78 to 0.99 with a median of 0.93, all significantly different from zero. For the second series of photographs, there was almost total concordance between the two sets of measurements performed by the same evaluator, with an intraclass correlation measurement of 0.99. The error median for the two Flexindex measurements was just 1 point, that is, 1 out of 20 measurements.

In terms of interobserver reliability, the experienced and inexperienced evaluators gave somewhat different scores. Interobserver reliability analysis revealed slightly lower levels of agreement than the intraobserver analysis, but it was almost always significant, with intraclass correlation coefficient medians of 0.88 and 0.83 for experienced and inexperienced evaluators, respectively, in the first series of photographs. In only two movements—ankle dorsiflexion and trunk flexion—were the intraclass correlation coefficients not significant for the inexperienced evaluator. The median of the Flexindex error for the two evaluators was similar, about two points. For the series of photographs of subjects undergoing the complete Flexitest, the intraclass correlation coefficients were quite high and similar for both evaluators at around 0.95, and there were no differences in the Flexindex averages. It was observed that each evaluator made an error of over 10% of the score points in only one subject (7% of the cases). Interestingly, the greatest number of errors was in the assessment of movement XI—trunk lateral flexion—which accounted for a little over 10% of the erroneous mea-

sures. In only 0.25% of the photos was there a 2-point error in the score on the dimensionless scale; three-quarters of those misjudgments were made by the inexperienced evaluator.

▾ **Conclusion:**

Flexitest assessment of joint range of motion enjoyed extremely high levels of intra- and interobserver reliability, not only for the first series of individual movements, but specifically for the Flexitest as a whole. These data suggest that the error margin tends to be slightly higher for extreme values and less-experienced evaluators. It is true that while photo analysis is a creative and original way to control the influence of the biological variability of the assessed subjects, it does not incorporate a potential source of error, which is the actual execution of the movements. On the other hand, the level of purposeful diversification of values in the first series of photos introduced a degree of difficulty quite beyond the everyday use of the Flexitest. Yet it can be assumed that *in vivo* reliability is somewhat less than the score obtained by using the photos, since most *in vivo* measurements tend to score a 2. We conclude that using previously standardized photos is a powerful tool for training future Flexitest users.

Trained evaluators show high intra- and inter-observer reliability for Flexitest measurements.

First Trial

Data originally from [7] and partially reanalyzed for this summary.

▾ **Rationale:**

A measurement tool should provide high indexes of reliability and validity, and ideally it should be easy for new evaluators to learn.

▾ **Objective:**

The objective of this study was to determine the margin of error in the first trial of new evaluators.

▾ **Sample:**

Our subjects were 23 physical education teachers and undergraduate students (7 men and 16 women) who attended a short course on flexibility theory and practice.

▼ **Methods:**

As part of a course on flexibility, Flexitest theory and methods were presented to the entire class. An expert evaluator demonstrated the method and then supervised as the students paired up to practice administering it with the maps (one student was to be tested, the other, to evaluate). Following this, a volunteer sample of the students (roughly 7% of the total class) was again subjected to the Flexitest by the instructors, who had no knowledge of the results found by the subject's partner in the previous test. The measurements assessed by the partner during training were later compared with the ones found by the experienced evaluators.

▼ **Results:**

The subjects showed the flexibility profiles expected for their age and gender, and the experienced evaluators determined the range of Flexindex scores to be 33 to 57 points, with a median of 50. There were no differences between the Flexindex results measured by experienced evaluators and students in their first trial. However, there were several individual mistakes that determined a rather modest intraclass correlation coefficient of 0.68 (see figure 7.2). Differences in the Flexindex scores given by expert and first-trial evaluators, which ranged from –9 to +11 points, occurred in nearly 40% of the cases. There was no clear tendency toward over- or underestimation globally or for each movement, except with movement XVI, for which the students tended to overestimate mobility. Of the 460 measurements taken, in only 3—all of movement XIX (shoulder lateral rotation)—did the error exceed 1 point on the Flexitest score scale. A more detailed analysis of the mistakes showed that they tended to occur more often for some movements (VIII, IX, XIX, and XX) than for others (I, II, IV, and XIII); however, there was not a significant predominance of mistakes in any specific part of the body.

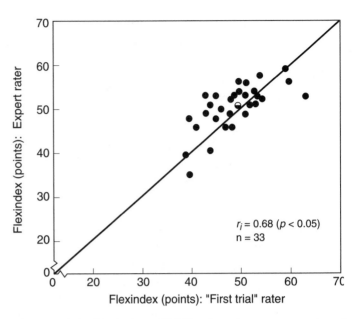

Figure 7.2 Flexindex reliability between an expert and "first-trial" evaluators. The half-filled circle represents three points that overlap.

▼ **Conclusion:**

At the end of a short course on flexibility theory and methods that had allotted some hours for Flexitest training, the students were already able to perform the test reasonably well and assigned global scores not significantly different on average from those assigned by experienced evaluators. However, when analyzed individually, the mistake may be important and eventually surpass the gain obtained after months of flexibility training. Some movements, especially shoulder rotation, are more difficult to learn and master than are others in the assessment technique. In general, however, mistakes of over 1 point in the assessment of a given movement were quite rare and limited to movement XIX. We have concluded that the error margin in an evaluator's first trial is relatively high and that therefore training in more than just one short course is necessary to ensure the reliable use of the Flexitest in practical and clinical situations.

> After a brief period of training, new evaluators can adequately assess Flexindex results, very rarely making 2-point errors in the evaluation of individual movements.

Concurrent Validity Studies

Data originally from [19, 27] and partially reanalyzed for this summary.

▾ **Rationale:**

The scientific characteristics of a good measurement tool include reliability and validity. Validity is the capability of measuring exactly what one intends to measure. Thus, a valid test's results reflect the variable being tested. A strategy frequently used to assess the validity of a new test procedure is to compare the results of the new procedure with those of other traditionally used methods. Doing so allows one to study the new procedure as a variable in a set of subjects. This strategy is called concurrent validity.

▾ **Objective:**

The objective of this study was to compare the results of different joint mobility tests in assessing a set of subjects chosen purposefully to reflect different degrees of flexibility.

▾ **Sample:**

Thirty subjects (16 men and boys and 14 women and girls) who had no locomotor system complaints were selected to take part in this voluntary study. The subjects' ages ranged from 2 to 54 years, with six children under 7 and four subjects over 50 years of age. Some were regularly active; others had a sedentary life.

▾ **Methods:**

Five flexibility testing procedures were used in this study, four of them of a dimensionless nature and one using a linear measurement scale. The linear method was the sit-and-reach test described by Wells and Dillon (1952), and it was administered according to the original methodology, except that the subjects were tested in bare feet. The dimensionless tests were the Rosenbloom sign (Rosenbloom et al. 1981), the toe-touch test, the Beighton-Hóran test (Beighton and Hóran 1970), and the Flexitest. All of them were performed according to their original methodology. All of the procedures were administered by a single evaluator in the same standardized sequence, with no previous warm-up or physical activity, and with the subjects unaware of the aim of the study. The average testing time was 10 minutes for each subject.

▾ **Results:**

The Rosenbloom sign results (the impossibility of superposing the faces and fingers of the hands) were negative for all individuals and therefore were excluded from subsequent statistical analysis. In the toe-touch test, only seven (23%) of the subjects were able to touch the tip of the big toe with the tip of the forefinger without bending the knees. For the Beighton-Hóran test, the extreme values of the scale were reached (0 and 9 points), resulting in a median of 3 and a clearly asymmetric distribution of the results, which showed a tendency to concentrate at the lower end of the scale. This did not happen in the sit-and-reach test, for which results ranged from −15 to +25 cm, with a 7-cm average and a fairly normal distribution of results. For the Flexitest, the results did not reach the ends of the scale and ranged from 35 to 68 points with a median of 49 points, displaying a Gaussian distribution. There was little association of the results of the Beighton-Hóran test with those of the sit-and-reach and toe-touch tests, with correlation coefficients of 0.26 and 0.40, respectively. These latter two tests were clearly related to each other ($r = 0.91$). The Flexitest, as represented by the Flexindex value, was significantly related to different degrees to all the tests, being in best accordance with the Beighton-Hóran test ($r = 0.81$).

▾ **Conclusion:**

Even though there was significant correlation between the results of the Flexitest (as represented by the Flexindex) and the other tests, which suggested the existence of concurrent validity, the degrees of association were relatively unassuming. This indicated that one could not with reasonable accuracy predict the results of any other test by using Flexitest results, or vice versa. The best association was with the Beighton-Hóran test, which assesses primarily ligamentous hyperlaxity. However, unlike in the Flexitest, floor and ceiling values are often found in the Beighton-Hóran test, thereby minimizing its discriminative power. In fact, 20% of the subjects in our sample scored 0 and a child achieved the maximum score, reducing the degree of association with the Flexitest. The sit-and-reach and toe-touch tests depend primarily on trunk flexion mobility, which is just one of the movements tested by the Flexitest. Considering the nature of movement specificity for both joints and movements, it is no wonder that correlations between the Flexitest (which measures 20 movements) and the other two tests (which measure just 1 movement) were only mild. This study thus corroborates Flexitest's validity while also pointing out its peculiar characteristic of the lack of floor and ceiling effects and its potentially higher degree of generalization in the assessment of a subject's overall flexibility.

> The Flexindex score is better correlated with Beighton-Hóran test scores (given for nine movements) than with sit-and-reach test results (one movement).

Strictly Methodological Studies

The development of a new test protocol requires considerable time and effort to ensure careful standardization of the procedures. While we were developing the Flexitest, we carried out several studies specifically related to its methodological aspects. Some of them are briefly reported in this section.

Lateral Dimorphism

Data originally from [6] and partially reanalyzed for this summary.

▼ Rationale:

For most human beings, one side of the body has a higher motor predominance than the other, a tendency that is called laterality. Most people are right-handed, but a small proportion—about 10%—have motor predominance on the left side. In sports, laterality plays an important role, whether it is the impulsion leg or the stronger arm, and often it is associated with clear morphological asymmetry. However, there is little information on the existence of this lateral dimorphism for flexibility.

▼ Objective:

The objective of this study was to identify the presence of lateral dimorphism in overall and specific flexibility in young and physically active adults of both genders.

▼ Sample:

Ninety-two physical education undergraduate students (46 of each gender), all apparently healthy and without relevant locomotor restraints, volunteered for this study. They were all physically active, with the typical age ranging from 18 to 22 years.

▼ Methods:

A single experienced Flexitest evaluator made the measurements. Each subject performed 16 movements bilaterally (movements IX, X, XVII, and XVIII were omitted).

▼ Results:

Flexindex values were in line with expectations for the students' age groups and genders, with median scores of 42 (range, 29-58) and 49 (range, 37-61) points for men and women, respectively. Bilateral Flexindex values were virtually identical and resulted in a very high intraclass correlation coefficient—r > 0.98—for both genders. A little more than half of the subjects, men and women alike, had a flexibility difference of at least 1 point between sides in a specific movement. In no case was there a difference of more than 1 point for the same movement performed on the two sides of the body. Of the 68 differences found (representing about 5% of all the bilateral measurements), 42% were in shoulder lateral rotation, with the higher score most often seen in the right side. Actually, almost two-thirds of the asymmetries were seen in shoulder joint movements. About 10% of the students presented with some degree of asymmetry in the maximum length of elbow movements. On the other hand, bilateral differences in ankle, knee, and wrist movements were rare, occurring in less than 3% of the subjects. Bilateral differences were equally frequent in all range of movement scores, i.e., with low or high ranges of motion.

▾ Conclusion:

Young and physically active adults, whether male or female, did not show high levels of lateral dimorphism in their flexibility. Differences were restricted to some movements only, particularly shoulder lateral rotation. Because this movement is critical in some basically unilateral sports, such as tennis, volleyball, and basketball, flexibility in the joint may be specifically affected by training and repeated motor activity in performing the movement extension. Therefore, in general, there is no reason to measure flexibility for both sides of the body. Even when there is some previous sport training, the right side tends to best represent the higher maximum range of passive joint movement. On the other hand, when something prevents or makes more difficult the measurement of the right side (i.e., a recent fracture immobilized in a cast), measuring the contralateral limb should reliably reflect the usual degree of mobility with no major risk of significantly compromising the Flexindex score.

▶ Concerning Lateral Dimorphism

- ▾ In general, flexibility is similar in the right and left sides of the body in both men and women
- ▾ In some subjects who have particular pathologies or in predominantly unilateral athletes, it may advisable to measure both sides

Circadian Variability

Data originally from [18] and partially reanalyzed for this summary.

▾ Rationale:

One of the most important biological rhythms is the circadian, which lasts approximately 24 hours. Over the circadian cycle, the body's hormonal and autonomic activity levels fluc-

tuate, changes that are reflected in sleep or wake status and other physiological conditions such as body temperature. Some ballet dancers have observed that at specific times of the day it is easier for them to perform some positions that require maximum ranges of motion. On the other hand, some rheumatic diseases impose movement restrictions early in the morning that progressively improve over the course of the day. There is little information available on flexibility fluctuations over the day in healthy subjects.

▾ Objective:

The objective of this study was to determine circadian patterns in the flexibility of young, healthy nonathletes.

▾ Sample:

Thirty recruits of the Brazilian Army volunteered to participate in the study. They were between 18 and 20 years of age and were undergoing military training, but did not excel in physical or sports performance.

▾ Methods:

Four complete sets of Flexitest measurements were taken by an experienced evaluator over a 24-hour period. The first set of measurements was made at noon, and thereafter they were repeated every six hours, with the subjects in the same order each time. The last set of measurements began at 6 A.M. on the following day. The recruits were kept in a military lodge over the time of the study, leaving the room only to eat and undergo flexibility measurement. Physical activity was restricted the day before and during the period of measurement.

▾ Results:

The Flexitest values for the 30 recruits were quite variable, ranging from 35 to 56 points and corresponding to percentiles 3 through 99 for the age group, with an average score of 43 points (percentile 45). In no Flexitest measurement (four sets × 30 subjects × 20 movements) was a score of 0 given, whereas a score of 4 was given for a little over 2% of the individual movement measurements. The mean results of the Flexitest showed limited variation (between 42.4 and 44.9 points), and there was a good correlation among them ($0.68 < r < 0.91$; $p < 0.01$), with no marked influence according to measurement time ($p = 0.26$). In another approach that considered the time of day when subjects had their best and worst scores, it was noted that higher values for flexibility were more common in the early evening and morning ($p < 0.01$). There was no correlation between Flexindex and the observed circadian variability. Analysis of individual data showed that Flexindex changes ranged between 2 and 11 points over the four measurements, with the median being 5 points, or about 11% of the Flexindex absolute value. The assessment of flexibility's circadian behavior for each movement showed that about half the scores were constant for the four measurements, and very seldom was variation more than 2 points (3% of the cases). Measurement variations were more common for hip and shoulder movements and rare for ankle dorsiflexion, knee and elbow extension, and wrist flexion and extension.

▼ Conclusion:

Our data demonstrated that the circadian cycle had a small impact on flexibility in young, healthy nonathletes, with slightly higher results seen in the early evening and morning despite a high degree of variability among subjects in this respect. Circadian variability of flexibility was typically low (11%), and could be somewhat masked by the margin of error of an experienced Flexitest evaluator (about 5%). This variability did not depend on the level of flexibility of the subject being tested and seemed to be caused primarily by changes in shoulder and hip movement mobility, the major joints that are primarily restricted by muscles and connective tissue. It is possible that in highly flexible individuals, such as dancers, the circadian cycle may play an important role. According to our data, it is advisable to set a fixed schedule for making repeated measurements in a single individual to avoid potential, though small, circadian interference with the results.

▶ Concerning Circadian Variability

▼ Overall flexibility changes are minimal.

▼ Shoulder and hip flexibility are the most influenced.

▼ A clear morning-afternoon-night pattern does not exist.

▼ For follow-up data analysis, standardize the time of measurement.

Observational Studies

After having assessed the reliability and concurrent validity and analyzed the most relevant methodological issues related to the Flexitest, we are now able to review some of our observational studies. Most of these studies analyzed the relationships between flexibility as assessed by the Flexitest and other morphofunctional variables.

Height-Weight Ratio and Somatotype

Data originally from [9] and from unpublished laboratory material.

▼ Rationale:

Kinanthropometric assessment involves aspects such as height and weight measurements, weight-height ratio, proportionality, body composition, muscle strength, and flexibility. A number of these elements are part of the physical fitness concept. The relationships among these different variables depend upon the characteristics of the population under investigation, their gender, and their age group, and data collected in a certain situation may not be duplicatable in another. This also holds true when specific groups, such as athletes of particular sports, are assessed. A more appropriate approach to kinanthropometric analysis may be to determine the levels of association among variables using data from a more heterogeneous sample population.

▼ Objective:

The objective of this study was to determine the association of height and weight measurements, height-weight ratio, and somatotype components with overall and specific flexibility.

▼ Sample:

Two hundred fifty-four subjects (116 men and boys and 138 women and girls) with a mean age of 48 (range, 13 to 82) in both gender groups were subjected to a thorough clinical-functional assessment in a specialized clinic. Subjects were retrospectively selected, and those who took part in competitive sports or for whom measurements were missing or incomplete were excluded.

▼ Methods:

Anthropometric measurements were made according to well-established methodologies. For the weight-height ratio, the Quetelet index, better known as the body mass index (BMI), was calculated. It is the ratio between the weight measured in kilograms and the square of the height measured in meters (kg/m^2). The somatotype was determined using the Heath-Carter anthropometric technique originally described in the 1960s. Flexibility was measured with the Flexitest.

▼ Results:

Measurements for height and weight and the height-weight ratio for men and women were respectively (mean ± standard deviation) 175 ± 6 and 161 ± 6 cm, 79 ± 14 and 67 ± 15 kg, and 26 ± 4 and 26 ± 7 kg/m^2. As expected for somatotype, women had higher values for endomorphy and ectomorphy and smaller scores for mesomorphy, with mean values of 5.5-5.1-0.7, while the corresponding male values were 4.5-5.5-1.3 ($p < 0.05$). Our sample had mean flexibility values in the 40th percentile for men and 63rd percentile for women at equivalent ages. Women were more flexible in all movements and joints and had higher overall flexibility: 46 ± 11 in women versus 34 ± 11 points on the Flexindex in men. There was some degree of association between height and overall flexibility, with correlation coefficients of 0.15 for men and 0.21 for women. By contrasting the 20 shortest and tallest subjects for each gender, it was noted that the tallest subjects were about 15% more flexible

($p = 0.13$ for men and $p = 0.02$ for women). Body weight and BMI were inversely related to flexibility, particularly in women ($r \geq -0.50$; $p < 0.01$). For other somatotype components, an inexpressive inverse relationship existed between mesomorphy and flexibility, but relative linearity, expressed by ectomorphy, was directly related to flexibility, slightly more so in females ($r = 0.59$) than in males ($r = 0.39$). Among predominantly ectomorphic subjects, the mean Flexindex percentiles were positively skewed as compared to the mean percentiles of the entire group, with 25% of the 12 ectomorphic women occupying Flexindex percentiles higher than 90 for their respective ages. Lower-limb joint motion, particularly of the hip, tended to be smaller the greater body weight, BMI, and endomorphy and mesomorphy values were.

As for movements, it was noted that hip adduction and abduction scores were inversely related to weight, BMI, and endomorphy values. Especially evident in women was an inverse relationship between mobility in five movements—knee flexion, hip extension and adduction, and shoulder posterior adduction and posterior extension—and endomorphy and mesomorphy values, as well as a direct association of these to relative linearity, with absolute correlation coefficient values ranging from 0.43 to 0.66 ($p < 0.01$). Shorter men had limited mobility in trunk movements compared to taller ones ($p < 0.05$).

▼ Conclusion:

There were some significant relationships between anthropometric measurements, somatotype, and body flexibility. Taller adults tended to be more flexible in some movements. Greater weight due to muscle or fat mass, especially in women, negatively affected body flexibility, particularly hip maximum range of motion. Relative hypomobility was typical of the somatotypic patterns of endomorphy and mesomorphy, whereas hypermobility was more prevalent in ectomorphic subjects, especially women. However, due to considerable superposition of data and relatively modest associations, it was not possible to infer the flexibility of a given subject based solely on anthropometric measurements or somatotype.

> Endomorphic and mesomorphic subjects typically are less flexible, while ectomorphs tend to be more flexible.

Handgrip Strength and Ability to Sit On and Rise From Floor

Data from unpublished laboratory material.

▾ Rationale:

Maintaining an adequate quality of life depends, at least in part, on maintaining satisfactory fitness levels. Today, there is growing concern about autonomy and the quality of life of the elderly. With aging, there is a tendency toward decreasing aerobic levels and flexibility and increasing sarcopenia (a progressive reduction of muscular mass that decreases the maximum levels of strength and power). When associated with hypomobility, sarcopenia tends to restrict autonomy in middle-aged and elderly subjects.

▾ Objective:

The objective of this study was to relate muscle strength, muscle power, and the ability to perform certain motor actions—sitting on and rising from the floor—to specific and overall flexibility.

▾ Sample:

We studied a sample of 254 middle-aged subjects (116 men and 138 women) balanced for gender and age. Subjects were selected from among people attending a specialized sports and exercise clinic for a detailed clinical-functional assessment. There was no special requirement for selection, but subjects who took part in competitive sport training were excluded.

▾ Methods:

Variable strength was assessed twice on each side by measuring bilateral handgrip with the arm fully extended and choosing the highest of the four measurements as representative for that subject. Absolute and relative-to-body-weight muscle power, in watts, were assessed by testing with progressively heavier loads in the half-row exercise, standing position, and measured with a tensiometer (Fitrodyne,

Bratislava, Slovakia). For the analysis of a motor action, we chose the sitting-and-rising test (Lira et al. 1999; Araújo 1999a), which is used to study dexterity in sitting on and rising from the floor. In this test, a score from 0 to 5 is given for each action, 1 point being deducted for using any type of support, such as the use of one of the hands or knees on the floor, and half a point being deducted if the subject displays any body imbalance when performing the movement.

▾ Results:

The distribution of Flexindex results ranged from 12 to 72 points, covering the whole spectrum of a typical population distribution, i.e., there were subjects representing percentiles from 1 to 99 for their respective gender and age group. Handgrip strength was significantly higher in men than in women (37.8 ± 7 versus 21.7 ± 5 kg; $p < 0.01$), but there was no significant dexterity difference in sitting on and rising from the floor. Just a portion of the sample (37%), 71 men and 23 women, undertook the maximum power test, with absolute and relative values being slightly higher in men ($p < 0.05$). There was a modest direct and identical relationship between handgrip strength and Flexindex scores in men and women, with correlation coefficients of 0.33 ($p < 0.01$). Sitting-and-rising performance is strongly influenced by overall flexibility, with correlation coefficients ranging from 0.58 to 0.73 ($p < 0.01$). Interestingly, the association between possessing the ability to sit on and rise from the floor is not especially better for any of the movements, with correlation coefficients being somewhat higher for lower-limb joint mobility. Although maximum absolute power is not related to flexibility ($r = 0.20$ for men and $r = -0.01$ for women), muscular power standardized by body weight shows a direct, but modest, association ($0.34 < r < 0.41$).

▾ Conclusion:

Contrary to what one might deduce from somatotype information (specifically, the inverse relationship between mesomorphy and flexibility in middle-aged subjects who do not take part in competitions or sports training), subjects with higher overall flexibility show above-average levels of muscle strength and power. These subjects are probably the youngest in the sample and also are more capable of performing motor movements that depend on these variables. These data are in accordance with the theory that with the passing of years, fitness levels decrease as a whole, not for just a single component.

> Ability to sit on and rise from the floor is favorably influenced by lower-limb flexibility.

Self-Perception

Data originally from [29] and partially reanalyzed for this summary.

▾ Rationale:

For initially learning and correcting problems in the execution of a given motor action, having knowledge of the benefits conferred by mastering a movement is extremely important. It is possible that for decision making in health and physical fitness, knowledge of results also plays an important role. Little is known about self-perception of physical fitness, and even less about self-perception of flexibility.

▾ Objective:

The objective of this study was to compare subjects' self-perception of flexibility with an expert's evaluation in adults practicing different levels of regular physical exercise and to determine if a discrepancy was associated with a specific level of body flexibility.

▾ Sample:

Fifty-two subjects (34 men and 18 women) ranging in age from 19 to 51 years were studied. Thirty were athletes, 19 were physically active, and 3 were sedentary. For the purpose of this study, athletes were defined as those who actively engaged in sports competition or who regularly had sports training, and the physically active were those who exercised at least three times a week.

▾ Methods:

The Flexitest procedure was explained and the evaluation maps were shown to all the subjects. Subjects were questioned about their current overall flexibility level according to a six-level ordinal scale (the corresponding percentiles are in parentheses): very poor (< P_{10}), poor (P_{10-25}), low average (P_{26-50}), high average (P_{51-75}), good (P_{76-90}), and excellent (> P_{90}). Percentiles were taken from the Flexitest data bank according to age and gender. Then the evaluator administered the Flexitest to the interviewee and the results of the subject's self-perception and evaluator's assessment were compared.

▾ Results:

There were no significant differences between the evaluator's measured and the subject's self-perceived Flexindex scores, whether in terms of ordinal classification or by comparison of average scores. Individually, however, disagreements were substantial, ranging from –25 to 28% of the effectively measured Flexindex. About a quarter of the subjects made substantial mistakes of over 10% in overestimating their flexibility, whereas 40% of the subjects underestimated their flexibility by less than 5%. A habitual physical exercise pattern had no influence on the right or wrong self-assessment of overall flexibility. The percentage of error in the self-perception assessment as a Flexindex value is related to the overall flexibility effectively measured ($r = 0.4$; $p = 0.002$).

▾ Conclusion:

In a generic way, self-perception of flexibility was found to be appropriate; however, on an individual basis, the error margin was excessive and can supplant eventual gains that may be conferred by specific training. Discrepancy between objective and subjective flexibility values seemed to be related to the overall level of flexibility and independent of participation in a regular exercise pattern. More-flexible people tend to underestimate their real flexibility levels and less-flexible people tend to overestimate their flexibility status, which ultimately implies a regression trend toward the mean.

> Self-perception of flexibility in adults tends to be reasonably accurate, although individual errors can be substantial.

Mitral Valve Prolapse

Data originally from [31] and partially reanalyzed for this summary.

▼ Rationale:

Mitral valve prolapse (MVP) is an abnormal, excessively lengthy, and redundant movement of one or more leaflets of the mitral valve. It is at least three to four times more frequent in women than in men, affecting 2 to 5% of women. Diagnosis is typically established by bidimensional echocardiography. It has been suspected that women with MVP have higher ligamentous laxity and often, higher overall flexibility.

▼ Objective:

The objective of this study was to assess if adult women with MVP have higher overall and specific joint movement flexibility.

▼ Sample:

One hundred twenty-five women having a mean age of 50 were retrospectively studied. Thirty-one of them had been diagnosed with MVP.

▼ Methods:

The Flexitest was applied by a single evaluator. Flexindex data for each of the movements and joints were compared in the groups with and without MVP. Because of the theoretical assumption that there would be difference, a unicaudal statistical approach was chosen.

▼ Results:

Flexindex results (48 versus 42 points) showed that women with MVP were about 15% more flexible on average than those who did not have this structural abnormality. None of the adult women studied exhibited overall hypermobility, i.e., a Flexindex score of over 70 points, although scores of over 55 points were at least three times more common in women with MVP. As expected, the flexibility evaluation revealed higher values in all joints for women with MVP; however, differences were not statistically significant for knee or trunk movements and were borderline for range of motion in the wrist.

As for individual movements assessed with the Flexitest, again there was a clear tendency toward higher values in the affected women, reaching statistical significance in 13 of the 20 movements. Interestingly, movements typically used in other flexibility assessments, such as knee extension, which is used to test hyperextension in the Beighton-Horán method, and trunk flexion, which resembles the sit-and-reach test, did not show statistical differences. For trunk flexion, most of the subjects in both groups scored a 2, and the means were practically identical (2.13 and 2.09, respectively, for MVP and non-MVP females). On the other hand, for elbow extension, there was a clear difference between the two groups, with means of 2.65 and 2.11, respectively, for MVP and non-MVP females. Moreover, scores of 0 or 1 for this movement were not found in women with MVP, but were found in at least 10% of women who did not have MVP, defining an excellent specificity. Another movement for which the absence of scores of 0 or 1 reveals high discriminative power is shoulder lateral rotation, with 0 versus 20%, respectively, occurrence rates in women with and without MVP. Scoring a 4 in the Flexitest was at least twice as common in adult women with MVP, but it also occurred in at least 5% of the movements of unaffected women. The prevalence of at least 3 scores of 4 from among those for the 20 movements was 25 and 10%, respectively, for the cases with and without MVP. Intermovement and intra-articular variability indexes did not differ between the two groups.

▼ Conclusion:

Although there was a tendency toward higher overall and specific flexibility in adult women with MVP, it was not always possible to identify them based exclusively on flexibility evaluation. The presence of important or relative hypomobility in elbow extension and in shoulder medial rotation seemed to exclude the possibility of MVP in adult women. On the other hand, trunk flexion, typically assessed with the sit-and-reach test, possessed the smallest discriminative power to identify the presence of MVP in female adults. The Flexitest, whether by 2 of its movements or by the whole set of 20 movements, may contribute to the physical examination and clinical assessment of adult women with MVP.

> Although overall hypermobility is a common finding in women with mitral valve prolapse, hypomobility in elbow extension and shoulder medial rotation practically exclude its presence.

Interventional Studies

To close this chapter, we present three studies in which the flexibility responses to different interventions are concisely described and the potential implications are discussed. These studies required test–retest Flexitest measurements and are likely useful for the reader interested in applying the method in this context.

Warm-Up Effect

Data originally from [11] and partially reanalyzed for this summary.

▾ Rationale:

It is typical for athletes and physically active subjects to start a training session with stretching exercises. For some athletes—gymnasts and dancers, for instance—these exercises are considered mandatory, taken quite seriously, and performed for a rather long period of time during each practice session. It is common sense that doing this prepares the body for the main and most important part of the training session and minimizes or reduces the chances of lesions of the locomotor system. It is also quite possible that stretching exercises promote a significant increase in flexibility, whether of a short-duration response or a long-term adaptation process. In the first case, it is fundamental to assess whether the warm-up affects the measurement of body flexibility, which would have relevant methodological implications for applying the Flexitest.

Standardization of a warm-up series or routine is likely to be feasible only in a highly controlled sports environment, such as the training sessions of a high-level competitive team. But the typical lonely sports enthusiast is likely to lack a systematic warm-up routine, whether it is because he does not know the techniques, does not value it, or does not prioritize so as to have the time to do it.

▾ Objective:

The objective of this study was to quantify the effects of active, nonstandardized, and self-planned warm-up on overall and specific flexibility in apparently healthy subjects of different ages and sports backgrounds. Differences caused by warm-up would then be related to age, initial flexibility level, and physical activity pattern.

▾ Sample:

A total of 109 volunteers (57 men and boys and 52 women and girls) between the ages of 6 and 35 years took part in this study. Physical activity patterns varied among the subjects and included children taking school physical education and swimming classes and physically active or sedentary adults.

▾ Methods:

A single experienced evaluator applied the Flexitest on the right side of all subjects before and within five minutes after an active, self-planned warm-up. The warm-up was independently performed for some 10 minutes after the subject was instructed to "warm up" the body joints through movement. In order to simulate the everyday conditions and allow results extrapolation to ordinary people who exercise, there was no control on the type, intensity, sequence, or magnitude of the joint range of motion reached during warm-up exercises. Proper descriptive and inferential statistical techniques were applied in the study and a significance level was established at 5% of probability.

▾ Results:

The warm-up fostered a significant increase (a median of 2 points) in the Flexindex value for all sample subgroups analyzed, ranging from −1 to +11 points. In 86 subjects (79%), flexibility increased with the warm-up, and in 37 of them (34%) this increase was over 10% of the initial Flexindex value, about 4 points, certainly more than an eventual error margin of the evaluator would be. All differences for individual movements were, as expected, of limited magnitude and represented by only 1 point on the Flexitest score scale. The positive impact on joint mobility conferred by warm-up was not evenly distributed among the 20 movements. The movements in which at least 15% of the subjects improved were: II, V, VIII, IX, X, XI, XVI, XIX, and XX; all joints except the knee, wrist, and elbow improved. All trunk flexibility movements benefited from the warm-up. On the other hand, there was no knee or elbow extension enhancement in any of the subjects (< 1%). As for the relationships between variables, it was noted that pre- and post-warm-up flexibility differences were not correlated with age-typical exercise behavior patterns or initial flexibility level.

▼ **Conclusion:**

Active warm-up, even when not standardized, positively influenced overall and specific flexibility, promoting significant mobility increases in trunk movements and in some movements of the hip, shoulder, and ankle. However, the gain was relatively small—some 5% of the initial value—and occurred in subjects with a higher degree of flexibility as well as in those with more limited overall mobility, thus suggesting that even highly flexible people such as gymnasts and dancers may achieve even higher mobility with warm-up. Other studies are necessary to define the warm-up routine that best promotes flexibility gains. Based on these data, it can also be concluded that it is necessary to control the variable of physical activity or warm-up in the hour prior to applying the Flexitest. Considering the broad range of flexibility gains with warm-up, it seems most appropriate to eliminate the variable by standardizing the application of Flexitest to prohibit a pre-evaluation warm-up.

> Just 10 minutes of self-planned, active warm-up improves flexibility mostly in trunk movements in both children and adults.

Ability to Perform Daily Activities

Data originally from [28] and partially reanalyzed for this summary.

▼ **Rationale:**

With the progressive increase in life expectancy, there is a substantial increment, both in absolute and relative figures, of the elderly population. Despite this very clear increase in the quantity of life, the quality of life issue has not been properly addressed. These subjects often present with chronic degenerative diseases that substantially affect their autonomy and quality of life. Plainly, adults and the elderly complain of an increasing degree of difficulty inherent in performing their everyday chores. It is possible, however, that staying physically active, and more specifically, regularly performing stretching exercises, may help to minimize these difficulties.

▼ **Objective:**

The objective of this study in adults was to relate general and specific flexibility gains in supervised physical exercise programs with eventual facilitation in performing daily physical exercises.

▼ **Sample:**

Twenty subjects (15 men and 5 women) were evaluated, most of them undergoing secondary prevention for coronary artery disease; their ages ranged from 38 to 76 years (57 ± 19 years). Subjects who fulfilled the following criteria were selected from the population of patients of a supervised physical exercise program:

▼ Flexibility assessment performed upon entering the program and within 3 to 18 months of training

▼ Regular participation in the program of three or more times a week, with no significant clinical intercurrences and with a compliance rate of over 75% during the period between the two flexibility evaluations

▼ Administration of the Flexitest by the same evaluator both times

▼ **Methods:**

As part of a broad routine of clinical and functional tests, the Flexitest was used to evaluate flexibility in the 20 subjects. Each subject was assessed and reassessed by the same evaluator, who was quite experienced in using the technique. In addition to performing these tests, the subjects were interviewed and answered a specific 11-item questionnaire to enable us to eventually assess any benefit derived from the frequent supervised physical exercise. The questionnaire sought to establish the degree of ease each subject reported for performing the following actions:

1. Going up and down steps
2. Getting in and out of a car
3. Tying shoelaces
4. Pedaling a bicycle
5. Walking on a treadmill
6. Crossing the legs while seated
7. Reaching the back to scratch or scrub when taking a shower
8. Reaching an object on top of a bookcase
9. Walking
10. Getting out of bed
11. Ducking

The relative level of ease for each of the 11 actions was determined on a visual analogical scale 10 cm long, with "very difficult" and "very easy" indicated at its left and right ends, respectively. Subjects were interviewed once in one session after months of training and were asked to mark on the scale their estimation of the level of difficulty before and after training. The distance in millimeters between the two marks was considered as representing the level of change due to training and could be positive, indicating improvement, or negative, showing worsening. This estimate was determined for each individual action and for the set as a whole. The supervised exercise program included about 30 minutes of an aerobic workout designed to achieve an individualized target heart rate determined according to the results of the initial cardiopulmonary exercise test, 8 to 10 muscle-strengthening exercises performed in two series of six to eight repetitions each, and about 10 minutes of active and passive stretching exercises customized for each subject to enhance mobility in those who were more restrained in the initial assessment.

▾ **Results:**

The supervised exercise program provoked an average 10% increase in overall flexibility and induced improvements in specific hip, trunk, and shoulder joint motions. Increments in hip extension and adduction, trunk lateral flexion, shoulder rotation, and back extension mobility were observed. There was also a significant enhancement in the ease of pedaling a bicycle, going up and down steps, and walking on the treadmill, and less impressive gains in crossing the legs and getting out of bed. The Flexindex gain was positive and significantly related to a higher level of ease in the overall performance of the everyday actions assessed ($r = 0.45$; $p < 0.04$). Furthermore, significant associations were also seen between: a) the extension of knee flexion and the ease of getting in and out of a car, and b) putting on and lacing the shoes and getting out of bed with the greater ease of increased trunk flexion mobility.

Despite these results, a further subdivision of the sample according to body weight showed that only for those who had a significant weight loss (i.e., 5% of original body weight) was there a major gain in flexibility; none of the four subjects who gained weight over the period of the study had any improvement in flexibility. In fact, there was an inverse and significant correlation between body weight and Flexindex variations ($r = -0.66$; $p < 0.05$). We also noted that those subjects who had lower initial levels of flexibility gained the most with the supervised training ($r = -0.65$; $p < 0.05$).

▾ **Conclusion:**

Adults submitting to a supervised exercise program improved their flexibility, especially of the major joints, which was reflected in the increasing degree of ease with which everyday actions were performed. However, there were important interindividual variations in the results that seemed to depend on the degree of initial flexibility and the presence of any concurrent changes in body weight. Because autonomy may be adversely affected by the aging process, it is possible that physical exercise programs with aerobic, muscle-strengthening, and stretching exercises may help to minimize or even to partially reverse these limitations on performing everyday chores, especially in individuals who manage to lose some weight and have low initial levels of flexibility.

> Subjects with poor levels of flexibility are the ones who display larger improvements in performing daily life activities after participating in a supervised exercise program that includes stretching exercises.

Supervised Exercise Program

Data from unpublished laboratory material.

▼ Rationale:

The regular practice of physical exercises brings about a number of positive physiological changes, leading to improvement in the most important aspects of physical fitness. However, it is clear that most of the physiological benefits of a regular fitness program are maintained only while it is continued; once interrupted, these benefits tend to be lost. Furthermore, the benefits are quite specific: muscular strength training does not focus on aerobic aspects of fitness, and a program based solely on aerobic exercises should not be expected to considerably improve body flexibility. Another major issue is related to the duration of follow-up. Most studies using physical exercise as an intervention last only weeks and rarely a few months, limiting the extrapolation of the information for a physically active individual over the long term. The flexibility benefits accrued during fitness programs have not been studied.

▼ Objective:

The objective of this study was to determine the specific and overall flexibility profile in adult subjects undergoing a supervised regular fitness program.

▼ Sample:

Fourteen adult subjects 61 ± 10 years of age with heart disease took part in a professionally supervised exercise program (typical staff to patient ratio of 1:3) that included aerobic,

muscle-strengthening, and flexibility exercises in sessions lasting from 60 to 90 minutes three to six times a week.

▼ Methods:

At the beginning of the program, subjects submitted to a detailed assessment, including the measurement of direct maximum oxygen uptake, muscle strength, and flexibility and the evaluation of work capacity using a ramp protocol in a cycle ergometer. Flexibility was assessed with the Flexitest. Based on the results, each subject attended an individualized exercise program that was prescribed and frequently adjusted to provide appropriate stimulus for improvement in all aspects of physical fitness. After an average period of 6 months (range, 3 to 14 months), the test routines were repeated by the same evaluator. In an attempt to avoid potential bias, this evaluator had no access to previous Flexitest results when reassessing a subject.

▼ Results:

On average, there was an increase of about 30% in Flexindex values, rising from 27.1 ± 2.1 to 34.3 ± 2.5 points (mean $\pm$ standard error of mean) ($p < 0.01$), with individual changes ranging from 0 to 14 points. Considering age group and gender, subjects in the initial assessment ranked on average in percentile 37 (range, 1 to 83); on reassessment, after the improvement in Flexindex scores, they moved to a mean percentile of 63 (range, 17 to 97). At first, 9 of the 14 subjects (70%) fell below the expected for their age group (percentile 50); after the training program, only 4 subjects (29%) had Flexindex scores lower than percentile 50 for their ages. There was not a significant relationship between initial values and their respective flexibility gains. In spite of the small sample size, it was possible to detect significant mobility improvements in 14 of the 20 Flexitest movements ($p < 0.05$), although no statistically significant differences were apparent for elbow, wrist, shoulder medial rotation, and trunk flexion movements. When analyzed by joint, it was noted that mobility improved in all joints ($p < 0.01$) except the wrist-elbow set of movements ($p = 0.22$). There was also a significant reduction in the distal–proximal variability index, from 2.05 ± 0.63 to 1.19 ± 0.05, with a trend toward body flexibility profiles more typical of younger subjects ($p = 0.09$).

▼ **Conclusion:**

A customized and well-controlled fitness program may provide significant overall and specific flexibility gains. Anatomical characteristics may explain small mobility gains for wrist and elbow joints. Relating the mean Flexindex score to the group's mean age, we have pre- and post-training percentile results of 35 and 63, respectively, reflecting the importance of intervention for improvement. Therefore, the

subjects' achievement of percentile 63, the average post-training Flexindex result for the age group 41-45, indicates that a supervised fitness program with customized stretching exercises undertaken for an average period of six months may allow subjects in their 60s to maintain the same flexibility standard they had when they were 20 years younger!

Cardiac patients who submitted to a few months of a customized active and passive stretching program improved their flexibility from below to above the average norms for age and gender.

Comparative Analysis of Testing Methods

At this point, after reading the other chapters, you may be interested in comparing the different ways to assess flexibility so you can pick the method that suits you the best. An ideal flexibility testing protocol should have the requisite characteristics shown in figure 8.1. This chapter will assist you in choosing among the main flexibility assessment methods by critically commenting on the benefits and drawbacks of each as revealed by the 18-criterion classification system presented in chapter 3 (see table 8.1). While there are a number of testing protocols in the different measurement systems (linear, angular, and dimensionless), we have selected the eight most relevant tests for this comparative analysis:

1. Flexitest
2. Leighton's technique
3. Goniometry
4. Beighton-Hóran
5. Nicholas test
6. Rosenbloom test
7. Cureton's test
8. Sit-and-reach test

The most-used linear test in the field of physical education and fitness is the *sit-and-reach*, which has different versions and adaptations (Jones et al. 1998; Holt, Pelham, and Burke 1999). One of the main reasons that this test is broadly used is that tables of performance criteria for different age groups and both genders are easily available, as well as the fact that it has been endorsed by renowned

scientific institutions (ACSM 2000). *Cureton's test*, widely used in the past, is being used less and less often, despite its meritorious inclusion of a number of movements. The sit-and-reach and Cureton's tests will serve as representatives of linear tests in the comparisons we make in this chapter.

For angular tests, the conventional *goniometry technique* is the gold standard and the method most frequently used by physical therapists, physiatrists, and orthopedic surgeons (AAOS 1965). An important variation is *Leighton's technique*, which uses a specific tool to measure the full movement arc without breaking it down into portions of flexion and extension (Leighton 1942, 1956). There are normative values available for evaluating most angles of movement arcs, although in some cases (for instance, trunk flexion) it is not possible to accurately align the masts of the goniometer to find an adequate central fulcrum. On the other hand, there is little information available about angular measurements in athletes or special groups such as ballet dancers. For comparison, we will address conventional goniometry and Leighton's technique.

An ideal flexibility testing method requires:

▼ High inter- and intra-tester reliability
▼ High safety (no significant risk of injury or death)
▼ Simplicity (no expensive devices, not time consuming)
▼ Good ease of training evaluators
▼ Availability of age- and gender-specific norms
▼ Specific movement and joint information and comparisons
▼ Availability of specific criteria for flexibility profile comparisons
▼ Distinction between active and passive measurements
▼ Calculation of an overall flexibility score with Gaussian distribution

Figure 8.1 Requisites for an ideal flexibility testing method.

Table 8.1 Comparison of Major Flexibility Testing Protocols According to 18-Criteria Classification System

No.	Criterion	Flexitest	Leighton	Goniometry	Beighton-Hóran	Nicholas	Rosen-bloom*	Cureton	Sit-and-Reach
Methodological									
1	Type of flexibility	Static	Static	Static	Static	Static	Static	Static	Static
2	Modes of execution	Passive	Active	Active	Active	Mixed	Active	Active	Active
3	Number of movements**	Multiple [20]	Multiple [30]	Multiple [>40]	Multiple [9]	Multiple [8]	Single	Multiple [4]	Single
4	Number of joints per test item	Single	Single	Single	Single	Composite	Single	Composite	Composite
5	Number of joint movements in a given test item	Single	Composite	Composite	Single	Composite	Single	Composite	Composite
6	Total number of joints measured	Multiple	Multiple	Multiple	Multiple	Multiple	Single	Multiple	Single
7	Total number of joint movements measured	Large	Large	Large	Large	Large	Small	Regular	Small
8	Global scoring capability	Yes	No	No	Yes	Yes	No	No	No
Operational									
9	Instruments required	None	Complex	Simple	None	None	None	Simple	Simple
10	Evaluation time	Medium	Long	Long	Medium	Medium	Short	Medium	Short
11	Feasibility	High	Medium	Medium	Very high	High	Very high	Medium	Medium
12	Measurement unit	Points	Degrees	Degrees	Points	Points	Yes or no	Inches	Inches
Scientific									
13	Reliability	High	High	High	Medium	Medium	Medium	High	High
14	Stability	High	High	High	High	High	High	High	High
15	Concurrent validity	High	High	High	High	Low	Low	Medium	Medium
16	Discriminatory power or sensitivity	Very high	Very high	Very high	Medium	Low	Low	Medium	Medium
17	Clinical and sports applicability	Very high	Very high	Very high	Low	Low	Low	Low	Medium
18	Data distribution characteristics	Parametrics	Parametrics	Parametrics	Non-parametrics	Non-parametrics	Non-parametrics	Parametrics	Parametrics

*Involves just small joints of finger, hand, and wrist areas.

**Trunk albeit reflecting the composite movements at all intervertebral disks is considered in our classification system, for sake of simplicity, as just one joint.

Most of the dimensionless tests are used for specific situations and purposes. Worth mentioning are the *Beighton-Hóran test* (Beighton and Hóran 1969), designed to identify hypermobility cases, the *Nicholas test*, which assesses football players before the season, and the *Rosenbloom test* (Rosenbloom et al. 1981), used for evaluating hand joint mobility to assess future risk of microangiopathy in diabetic teenagers. The Flexitest is a dimensionless test for medical and sports purposes (Araújo 1986, 2001). These four valuable dimensionless tests were selected for our comparison of linear, angular, and dimensionless methods.

Methodological Criteria

Before we begin, we should recognize that our comparison of the selected tests is limited to measurements of static flexibility, because the assessment of dynamic flexibility is mostly confined to special research settings and sophisticated labs.

The tests differ in the details of their performance: Some collect measures of active flexibility, others compile measures of passive flexibility, and yet others mix active and passive measurements, such as the Nicholas test, and are classified as mixed tests. Some of the range of motion measurement techniques, such as goniometry, may be applied in either a passive or active way, although the passive form is used less commonly because it typically requires two evaluators. This aspect is quite important for distinguishing among flexibility tests.

When assessing a specific biological variable, such as flexibility, it is best if the outcome is not tainted or influenced by other variables. For specifically measuring variable flexibility, passive as opposed to active measurement is preferred because muscular strength and motor coordination may play an important role in the latter method. This is clearly seen in the knee flexion movement performed with the subject standing. If the subject performs the movement only by contracting the flexor muscles, the anatomical site of the muscles significantly limits the range of motion. On the other hand, if the movement is repeated with the subject using her hands to pull the leg against the thigh, the range of motion is considerably larger. In fact, in passive knee flexion the posterior portions of the leg and thigh often can be superposed, demonstrating that the active range of motion was limited by the imposition of the contracting muscles, not by poor flexibility. An even more striking example of the difference between active and passive flexibility is the person with paraplegia who cannot voluntarily flex his hip but has a normal range of motion on passive hip flexion.

Two questions often posed about passive flexibility assessment are, How much effort should the evaluator apply, and How will one know when passive mobility has reached its limit? Good muscle strength is needed to apply passive methods, especially for performing specific movements in very strong or obese subjects. However, with technical skill and the typical strength of a young woman of average size, it is possible for an evaluator to apply Flexitest and the other passive methods with no difficulty. In the beginning of nearly all joint movements, the performance is quite easy, but it becomes very difficult when the maximum range of motion is approached. The extreme positions of most movements, when additional force is employed, may cause sudden discomfort in the test subject. The definition of flexibility that we presented in chapter 1 may be the only one that explicitly states that flexibility should be measured at physiological limits, meaning that no injury should be caused. Even when other definitions do not specifically address this concern, it seems clear that flexibility measurements should be taken without risking harm. Accordingly, Flexitest administration should be performed carefully and seek to reach the maximum physiological amplitude of a given movement while obviously respecting muscle, tendon, and joint stiffness in the subject being tested. This is also valid for other methods of passive evaluation.

It seems likely that the possibility of an accident or injury occurring during flexibility testing would be higher for active than for passive methods. A health professional would be less likely to injure a subject when performing a movement than a subject performing the sit-and-reach test would be to injure himself by swinging his trunk forward and overstretching the posterior thigh muscles to achieve an optimal result. Therefore, we believe passive flexibility assessment should be the mode of preference, with active or mixed measurements being left for very specific cases.

To evaluate flexibility, one may assess one or more movements. Since Harris' classic studies (1969a), it has been known that joint mobility is a very specific feature that may vary according to each movement or joint. The five methodological criteria we present next take flexibility specificity into consideration.

The first criterion relates to the number of movements in the test. While some tests measure mobility by means of a single body movement (for instance, the sit-and-reach test), others combine a number of movement measurements. We also consider the number of joints and the total number of joint movements that were effectively measured. One can stratify methods into groups of "small," "regular," and "large" collections of data on the basis of the number of single or repeated joint measurements taken and the total number of joint movements. For instance, the Beighton-Hóran test evaluates 9 movements, the Flexitest measures 20 movements (or 36, if both sides of the body are evaluated), and goniometry assesses over 40 angular measurements.

When more than one joint range of motion is evaluated in a test's movement, make a note of the number of joints and joint movements that are evaluated. A test is said to be of a simple design when only one joint movement is evaluated for each test movement and of a complex design for movement or joint when a test movement considers more than one movement or joint in assessing the score. For instance, in the sit-and-reach test, we assess a single movement whose outcome depends on the mobility of a number of joints and movements (ankle dorsiflexion, knee extension, trunk flexion, elbow extension, and others), so it is a test for both movement and joint. Another example is angular tests that combine elbow flexion and extension mobility to assess the elbow movement arc with Leighton's flexometer; this test is thus complex for movement classification and simple for joint.

With respect to flexibility specificity, whenever possible, measure all of the different movements of the principal joints. Tests that measure a single movement or joint's flexibility and those that combine several joint movements in one or a few test movements do not identify specificity and their evaluation is quite limited, especially for classifying a joint movement according to its amplitude to establish a medical diagnosis, define a favorable sport standard, or prescribe physical exercises. For these purposes, the sit-and-reach test is least suitable, whereas Flexitest, by measuring at least seven joints and 20 movements, is best suited for most professional health and sport purposes.

Most research indicates that children are more flexible than adults and that among adults of the same age group, women tend to be more flexible than men. Such statements show that flexibility, despite presenting a high degree of specificity, is a generic feature of both health- and performance-related physical fitness. Test outcomes do not always effectively reflect the measured flexibility, and tests may be classified according to whether or not they provide a final or consolidated global score. In single-movement tests such as the sit-and-reach test, the measurement itself is a global final score; however, its interpretation is quite limited due to a number of intervening variables (Shephard, Berridge, and Montelpare 1990). In tests that include a number of movements, all of the movements should be measured on the same scale and equal results should have equal meaning. This problem is, unfortunately, one of goniometry's (and Leighton's technique's) major limitations, because a 90° movement arc for hip flexion-extension reflects limited mobility, but in the ankle, it reflects hypermobility. Thus, goniometric measurement can be made of each movement's angle, but the results cannot be grouped to arrive at a global score. Flexitest's measurement of 20 movements on a single scale gives each score the same technical and statistical meaning and allows all the scores to be added to arrive at an overall flexibility score, called the Flexindex. In the Flexitest scoring system, the larger joints such as the shoulder and hip most influence the final result by being tested in five (25%) and four (20%) of the movements, respectively. This does not happen in the Beighton-Hóran test, which also can provide a global score that ranges from 0 to 9 points, but does not measure either shoulder or hip mobility.

Operational Criteria

We have discussed the first eight methodological criteria of the flexibility test classification system, so now we will move on to operational issues. Interestingly, despite the recognition flexibility has accrued as an important feature in physical fitness assessment and exercise prescription, little is done to promote this issue. In most professional settings, time, personnel, and materials needed to assess physical fitness and its main variables are scarce, and the shortage is even more marked in relation to flexibility. One of the clearer limitations to implementing a test method in a professional setting is a requirement for expensive and sophisticated equipment, like that needed to directly measure pulmonary ventilation and oxygen uptake during exercise. The equipment used for flexibility tests is less sophisticated, with the exception of electronics used in electrogoniometry or dynamic flexibility measurements. Flexibility tests can be classified according to the level of sophistication the equipment requires, with Leighton's technique and its flexometer being the most complex, followed by different types of plastic or metal goniometers and the standard sit-and-reach bench, and last, comparative charts used for evaluating flexibility with the tests of Beighton-Hóran, Nicholas, and Flexitest.

Another critical aspect of choosing a test to use is the degree of difficulty involved in learning and becoming skilled at its application and interpretation. Most flexibility evaluation tests are not effectively learned in regular undergraduate and graduate courses. Extra courses or workshops can be helpful, but more often, the new evaluator learns it herself by associating with a practitioner who knows the technique or studying technical and scientific textbooks like the one you are reading right now. Although some tests are quite simple (for instance, the Rosenbloom or sit-and-reach tests), others require some specific training (like goniometry), and a handful of them require an even longer time to learn because of the equipment that must be used or the technique itself (like Leighton's technique). In this respect, Flexitest is moderately difficult to learn because studying the method is required, as is studying the evaluation charts to focus on the relative positions of the evaluator and subject and the proper angle of visualization for an assessment to be made. However, by reading chapter 4 of this book and practicing with the self-training methods presented in chapter 5, the professional will be able to start assessing subjects almost immediately. With the support of the chapter 5 reference data and the statistical resources presented in chapter 6, the reader will be able to use the Flexitest effectively and efficiently.

One of the major operational limitations for flexibility evaluation is the amount of time it takes. Flexibility tests should take as little time as possible to optimize the evaluator's use of time. Some tests take less than 1 minute to perform in a previously prepared setting, such as the sit-and-reach. Other tests, among them the Flexitest, require between 2 and 5 minutes to perform, and a few tests, such as Leighton's technique and goniometry, can take more than 10 minutes, depending on how many movements are measured. Flexitest has a favorable time-use profile, as it measures 20 movements in less than 5 minutes.

In addition, tests must be performed under the proper environmental conditions, such as in an appropriate room with good illumination. All eight of the tests selected for comparison can be easily or fairly easily executed, with the hardest to perform being Leighton's technique and the sit-and-reach test and the easiest being the simple dimensionless tests like Beighton-Hóran and Rosenbloom. The ease of administering the Flexitest is intermediate, and it can be applied in almost any setting, from a doctor's office to an evaluation room, a game court, a fitness room, or even in the pool area.

An alternative way to distinguish among and classify flexibility tests that is based on operational criteria is to categorize them by the measurement unit employed, whether linear, angular, or dimensionless. Linear tests are always quantified with linear units such as centimeters and inches. Angular test results are expressed in degrees. Although there are no formal units for expressing the results of dimensionless tests, they can be quantified in points\scores (ratio and interval scales), progressive ratings (ordinal scales), or yes-or-no answers (nominal scales).

Scientific Criteria

The last set of criteria to be addressed in this chapter relates to the scientific aspects of the tests and addresses their reliability, stability, validity, sensitivity, applicability, and the nature of their results distribution.

Reviewing the meanings of these terms, a reliable test is one in which results achieved in controlled situations tend to be identical, i.e., with minimal or no variability. Reliability allows results to be compared before and after an intervention (for instance, a training program) or between subjects in such a way that one can understand the meaning of any differences that are found. In general, flexibility tests are highly reliable, with the exception of dimensionless tests like Beighton-Hóran and Rosenbloom, which offer a small range of scores. Flexitest is no exception, as it presents highly reliable results for inter- and intraobserver scores—similar to, and better than, other methods. Reliability is enhanced when the application and interpretation methodologies are carefully followed and the evaluator is skilled enough to identify minor differences. For instance, any of us could see which was the tallest of two subjects standing side by side, even if the difference in their height was just 0.5 inch, or less than 1%. We have conducted pilot studies that have shown that an angular difference of over 3% between two linear segments is correctly identified more than 95% of the time, and a margin of error of under 5% is less than the margin of error typically accepted in goniometric measurements. This visual ability to discern differences accounts for the high reliability of passive joint mobility scores assessed by visual comparison with Flexitest charts.

The main flexibility assessment methods are highly stable and present consistent results as long as they are carried out in controlled settings. The two main factors that affect flexibility test stability are pain perception or tolerance and the performance of physical exercise immediately before the measurement. Pain sensitivity and tolerance varies from one subject to another, and from one moment to another in the same subject, so in the presence of a recent injury, joint mobility may exhibit little stability according to the subject's sensibility and tolerance of pain. Recent research has pointed out that even a single stretch may affect the subject's tolerance of the following stretch (Magnusson 1998). Intense physical exercise—especially if it significantly raises body temperature—performed prior to testing may considerably reduce the stability of a flexibility test. Therefore, we recommend that flexibility measurements for Flexitest or any other test be made with no prior physical exercise or warm-up and no repeat trials performed with the aim of achieving higher stability of the results. Linear tests such as the sit-and-reach or Cureton's test may exhibit impaired stability when various consecutive measurements are made, which may be responsible for some gains eventually reported by teachers or instructors in these many measurements.

The validity of a measurement tool is one of the most important scientific criteria to be considered when choosing a test method. A valid and stable test is of no use if its measurements have no meaning or cannot be associated with the variable it was intended to measure and assess. Unfortunately, there is no gold standard available for comparatively determining the validity of flexibility tests; one can only compare concurrent validity by assessing the association among two or more tests. Another approach is to validate the test against reference criteria, such as low-back pain incidence, risk for falls in elderly people, or a specific sport performance (Jackson and Baker 1986).

More specific flexibility tests tend to be less valid, whereas more general ones that involve a higher number of movements exhibit higher concurrent validity. Results of angular tests, the Beighton-Hóran test, and the Flexitest are favorably correlated, which shows high validity; the same does not happen with the sit-and-reach test (Shephard, Berridge, and Montelpare 1990; Cornbleet and Woolsey 1996; Jackson et al. 1998; Chung and Yuen 1999), Cureton's test (Tully and Stillman 1997), the Nicholas test, or the Rosenbloom test. These latter tests, therefore, should be applied in more specific conditions, as when, for instance, one wants to assess whether a child is able to touch the toes with the tips of the fingers by bending just at the trunk, not the knee. More broadly valid tests, such as the Flexitest and goniometry, may be applied in more generic situations.

An important scientific feature is the sensitivity or discriminatory power of a test. A sensitive test is one whose results allow an evaluator to correctly identify a change in the condition of the subject. This is very important in the physical fitness setting when one wishes to measure improvement derived from a specific flexibility program or losses due to aging or a disease process. Tests that include floor or ceiling effects, i.e., whose minimum or maximum results are reached by a significant number of subjects, have limited sensitivity. For example, the minimum score, 0, in the Beighton-Hóran test (revealing a lack of hypermobility in the nine tested movements) is a common finding in middle-aged adults. Because there is no lower score, this test does not detect flexibility losses that take place over the years. Data obtained with the sit-and-reach test, probably as a result of anthropometric discrepancies among its subjects, has suggested that maximum body flexibility levels in pubertal children, not in younger children, which is not commonsensical and is contrary to evidence gathered by other methods. On the other hand, the Flexitest has been shown to be quite sensitive in a number of situations, correctly identifying differences between men and women, changes caused by aging, and even the effects of short-term stretching training performed by previously sedentary or limited-flexibility subjects. Flex-itest results shown in percentile curves are scientifically feasible and coherent, as opposed to what occurs with linear tests.

The applicability of a test for medical assessment or sports evaluation relates to its potential use in the specific setting being considered. While some tests have a broad range of application purposes, others tend to be restricted to very specific situations (for instance, the Rosenbloom test). The Beighton-Hóran test is very useful for assessing hypermobility, especially in children and youngsters, but it is not applicable in older subjects or for selecting athletes for different sports. The Nicholas and Cureton's tests focus on very specific groups, limiting their applicability. Even though the sit-and-reach test has had a historical role in stimulating physical activity and improvements in physical fitness of U.S. students, its applicability is probably quite restricted, and even attempts to relate poor performance in this test with back pain failed (Jackson et al. 1998). Goniometry lends itself to more general application, and it is probably the test of choice when one needs very accurate flexibility measurements for a given movement so that clinical decisions can be made (for instance, for assessment of knee flexion after reconstruction of the anterior cruciate ligament). The Flexitest has a broad scoring system that covers all mobility ranges from hypo- to hypermobility, movement by movement, in seven major body joints. The main movements and joints involved in almost all major body motions are included in the Flexitest. Furthermore, the method offers the unique feature of a consolidated global score with no floor or ceiling effect, thus considerably enhancing the applicability of the method. A unique feature of the Flexitest is the possibility of analyzing the pattern of flexibility variability among the movements and joints and between the body segments. The fact that the Flexitest offers a broad database and age and gender percentile curves based on data from almost 3,000 people of both genders, nonathletes and athletes, ranging in age from 5 to 88 years, further encourages its use for medical and sports purposes, because it probably has the broadest applicability of all of the flexibility tests.

The results distribution characteristics for statistical analysis are also of importance in assessing testing methods. Data may be distributed according to a normal curve and then either have mathematically known parameters or fail to meet distribution standards and are termed nonparametric. Linear and angular tests present results very close to the Gaussian curve and typically are considered parametric. On the other hand, dimensionless tests, due to the features of their measuring scales, generally are considered to be nonparametric. The Flexitest, the exception to this rule, is the only dimensionless test with intentionally parametric distribution of results for each movement and for the set of movements. Therefore, statistical analysis of its results may make use of more powerful statistical resources and procedures.

Conclusion

When an athlete or ordinary person is injured, extreme mobility reduction of one or more joint movements often results. Similarly, at a much slower pace over the years, mobility also decreases due to the aging process. In both cases, the subject recognizes the reduction in mobility, despite not being able to accurately quantify it; he only knows his maximum range of motion or flexibility is decreasing. A health professional may confirm this finding by engaging in an abridged physical examination, using mainly her vision and clinical expertise to judge whether the ranges of motion are normal or not. While this strategy is practical, the lack of a formal test protocol with established measurements and evaluation criteria implies results of low reliability and validity, impairing diagnosis, the establishment of proper intervention, and the long-term following of results. It is within this broader sense of health-related physical fitness that the need for measuring and evaluating flexibility fits.

Flexitest Case Studies

9

Practitioners and students who are interested in flexibility become excited as they realize the shift in focus that has been taking place over the past 20 years. Originally, research focused on flexibility in competitive athletes; now it is relevant for all individuals, from children to the elderly, from the untrained to the highly trained, from the healthy to the physically disabled. The study of flexibility brings about new possibilities and responsibilities for all those who advise in or supervise physical activity. Different from other health-related physical fitness variables such as muscle strength and maximal aerobic power, flexibility is not a variable for which larger values are potentially beneficial. The idea is not to make people as flexible as possible, but to identify their current flexibility level, their individual needs and ambitions and how to meet them, and in particular how to monitor and evaluate the process. If increasing the aerobic condition of an individual by 10% requires many hours of exercising, just a few minutes of exercise performed in two or three short sessions a week may effectively improve overall flexibility by 20 to 50%.

In part I of this book, you were informed of the most current information about flexibility and your knowledge of different joint mobility tests was reviewed and broadened. In part II, the Flexitest methodology was presented, practice materials were explained, and you learned about the different ways to analyze the results, including the use of Flexindex percentiles, specific evaluation of movements and joints, and variability indexes, a unique Flexitest feature that makes the test different from all other flexibility assessment tools. Part III began with brief descriptions of 13 research studies that used the Flexitest to respond to relevant and practical issues. In addition, we presented a comparative critical analysis of the principal joint mobility assessment methods—linear, angular, and dimensional—explaining the pros and cons of each type of application.

This last section of our book presents eight examples of interpreting Flexitest results. We have selected a variety of actual cases from our professional practice to present (see tables 9.1 through 9.8). Each case is presented in four parts:

1. A brief and focused description of the subject's history and physical examination
2. Scores and indexes provided by Flexitest
3. Descriptive analysis of the findings
4. Clinical and sports interpretation of the Flexitest results, including our recommendations for that subject

By presenting these cases, our intention is to further detail our experience with the Flexitest. By analyzing actual cases, the practical and relevant information that the flexibility assessment by the Flexitest provides—single movements, joints, variability indexes and Flexindex—will become apparent.

Case Study 1

Case Study 1 is an 18-year-old male tennis player with juvenile Association of Tennis Professionals (ATP) ranking who is 173 cm tall and weighs 61.5 kg. He has been seen four times over the last seven years in our Sports and Exercise Medicine Clinic. When he was last assessed, he was training some four to five hours a day, practicing tennis, aerobics, and strength training. The Flexitest data are presented in table 9.1.

Table 9.1										Case Study 1
Movement	**I**	**II**	**III**	**IV**	**V**	**VI**	**VII**	**VIII**	**IX**	**X**
Movement score	2	2	2	2	2	3	3	2	3	3
Movement	**XI**	**XII**	**XIII**	**XIV**	**XV**	**XVI**	**XVII**	**XVIII**	**XIX**	**XX**
Movement score	3	2	2	2	2	3	3	2	4	1

Joint	Standardized score				
Ankle	40				
Knee	40				
Hip	50				
Trunk	60				
Wrist	40				
Elbow	40				
Shoulder	52				

Variability index	Result				
IMVI	0.66				
IJVI	0.37				
FEVI	0.93				
BSVI	0.96				
DPVI	0.78				

Flexindex	Score	Percentile
	48	83

The Flexindex score is 48 points, corresponding to percentile 83 for age and gender and indicating good overall mobility, probably higher than that of many tennis players of his age. Variability indexes point to a heterogeneous body mobility pattern, with higher flexibility for proximal and extension movements. Analysis per joint shows higher flexibility for trunk, shoulder, and hip. For individual movements, there is above-average mobility for movements of the trunk and posterior adduction of the shoulder, an exceptional range of motion for shoulder lateral rotation, and limited medial rotation for the same shoulder.

These results are in accordance with a more aggressive style of play, with few ball changes from the end of the court and a preference to approach the net and finish the point. In practically all the more relevant joints and movements for tennis, he shows an above-average range of motion, making easier the movements close to the net, volleys, smashes, and serves. He does not exhibit restrained wrist extension values, which are typical of adult tennis players, probably due to the need to firmly hold the racquet in order to avoid vibration. We recommend, however, that he try to increase mobility in the ankle dorsiflexion and in hip flexion and extension, which could make it easier to reach those short balls close to the net and low balls in general.

Case Study 2

Case Study 2 is a 22-year-old female beach volleyball player who weighs 63.5 kg and is 175 cm tall. She was assessed in the Sports and Exercise Medicine Clinic just before taking part in the Olympic Games at Atlanta, Georgia, at which her team won a medal. She was in excellent physical shape when the assessment was made and had no medical complaints in the locomotor system. The Flexitest data are presented in table 9.2.

She scored only 44 points on Flexindex, corresponding to a percentile of 17, which reflects a quite low overall mobility. Variability indexes indicate a homogeneous flexibility profile, with low interjoint and intermovement variability. Analysis per joint shows a higher range of motion for the ankle, trunk, and shoulder. Only four movements received scores of 3.

These results are somewhat surprising for an Olympic medal winner, and indicate that an excellent performance in beach volleyball may be achieved with an overall flexibility level lower than the average for the reference population. When observing the athlete's playing style, one sees that her aerobic condition and muscular power are certainly above average,

Table 9.2										Case Study 2
Movement	**I**	**II**	**III**	**IV**	**V**	**VI**	**VII**	**VIII**	**IX**	**X**
Movement score	2	3	2	2	2	2	2	2	2	2
Movement	**XI**	**XII**	**XIII**	**XIV**	**XV**	**XVI**	**XVII**	**XVIII**	**XIX**	**XX**
Movement score	3	2	2	2	2	2	3	2	3	1

Joint	Standardized score	
Ankle	50	
Knee	40	
Hip	40	
Trunk	47	
Wrist	40	
Elbow	40	
Shoulder	48	

Variability index	Result	
IMVI	0.40	
IJVI	0.21	
FEVI	1.08	
BSVI	0.96	
DPVI	0.96	

Flexindex	Score	Percentile
	44	17

her technical skills are excellent in all volleyball fundamentals, and she has extremely regular performance. She does not stand out in terms of exceptional defenses, for which higher flexibility levels are more likely required. Higher ankle, trunk, and shoulder motion favor her technical skill, especially when attacking and blocking. Dorsiflexibility of the ankle, just about average and below average for age and gender, is a common finding in beach volleyball players and may be related to an advantageous mechanical transfer for jumping when there is less elasticity in this area. She could benefit if her flexibility were enhanced, especially in hip movements and shoulder posterior adduction, which are useful for volleyball.

Case Study 3

Case Study 3 is a 26-year-old man with a height of 171 cm and weight of 70 kg who practices high-intensity muscle-strengthening exercises. He shows a condition typical of muscular dimorphism, with a fully contracted arm girth of 38.5 cm, whereas for the leg it is only 35.4 cm. He denies the use of hormones, but confirms use of nutritional supplements. He was referred to us by another physician for consultation and advice, because his lab results revealed creatine phosphokinase levels many times higher than the normal range. The Flexitest data are presented in table 9.3.

Table 9.3										Case Study 3
Movement	**I**	**II**	**III**	**IV**	**V**	**VI**	**VII**	**VIII**	**IX**	**X**
Movement score	1	2	2	2	3	3	2	2	2	2
Movement	**XI**	**XII**	**XIII**	**XIV**	**XV**	**XVI**	**XVII**	**XVIII**	**XIX**	**XX**
Movement score	2	2	0	2	1	2	2	1	2	2

Joint	Standardized score	
Ankle	30	
Knee	40	
Hip	50	
Trunk	40	
Wrist	20	
Elbow	30	
Shoulder	36	

Variability index	Result	
IMVI	0.65	
IJVI	0.40	
FEVI	1.44	
BSVI	1.37	
DPVI	0.71	

Flexindex	Score	Percentile
	37	17

He scored 37 on Flexindex, corresponding to a percentile of 17, a quite low overall mobility. Some of the variability indexes show abnormal values, with a predominance of mobility in the lower limbs in extension and proximal movements. Analysis per joint indicates that wrist and elbow mobility are quite reduced. In terms of individual movements, there is some hypomobility in ankle dorsiflexion, elbow extension, and shoulder posterior extension, and a significant restraint of wrist extension.

These results show an abnormal body flexibility pattern with discrete overall hypomobility and significant distortions in the flexibility of some movements and joints, probably due to an unbalanced and inadequate muscle-strengthening workout program. We recommend that his current muscle-strengthening workout program and use of nutritional supplements be promptly reviewed and stretching exercises be immediately implemented to correct or lessen the problems found (including his position when walking, with semiflexed elbows). This should be done immediately, because wrist and elbow mobility are hard to improve when one gets older.

Case Study 4

Case Study 4 is a 41-year-old man with plurimetabolic syndrome (high blood pressure, dyslipidemia, and obesity) who weighs 120 kg

Table 9.4										Case Study 4
Movement	I	II	III	IV	V	VI	VII	VIII	IX	X
Movement score	0	2	2	3	1	1	0	1	1	2
Movement	XI	XII	XIII	XIV	XV	XVI	XVII	XVIII	XIX	XX
Movement score	2	2	2	2	2	2	1	1	2	1

Joint	Standardized score	
Ankle	20	
Knee	50	
Hip	15	
Trunk	33	
Wrist	40	
Elbow	40	
Shoulder	28	
Variability index	**Result**	
IMVI	0.74	
IJVI	0.57	
FEVI	1.00	
BSVI2	0.75	
DPVI	1.69	
Flexindex	**Score**	**Percentile**
	30	20

and measures 182 cm tall. He was referred to us by his attending physician for admission into our supervised exercise program. He has no complaints concerning his locomotor system that could interfere with flexibility measurement. The Flexitest data are presented in table 9.4.

Flexindex results show an overall hypomobility pattern matching a percentile of 20. There is high variability in the mobility of the different joints and flexibility predominance in the upper limbs and distal movements. The hip and ankle joints exhibit limited mobility. There is significant mobility restraint in ankle dorsiflexion and hip adduction, and a number of other movements have reduced ranges of motion. Relative hypermobility of knee extension is of interest because it could place excessive stress on the joint while he is walking or descending stairs, as a function of his highly overweight condition.

These results show that overall and specific mobility are adversely affected by overweight (body fat and muscular mass), leading a relatively young person to present with the flexibility pattern of an elderly person. Significant difficulty in performing hip adduction is quite common in individuals with high levels of body fat and muscular mass. The reduced amplitude of ankle dorsiflexion tends to compromise gait, regaining balance in almost-fall cases, and standing from a chair or bed. The central obesity pattern makes trunk flexion too difficult. We recommend a program of stretching exercises in association with a broader program of general exercise and a diet. One should try to reduce knee hyperextension through specific muscular strengthening exercises, to minimize the chance of falls or torsions.

Case Study 5

Case Study 5 is a 52-year-old woman weighing 67 kg and measuring 169 cm who has mitral valve prolapse and complains of palpitations related to exercise. She came for advice on physical activity and reports having been very active until two years ago, when she decreased the intensity of her regular physical exercise. She also reports pain in the knees on climbing stairs. The Flexitest data are presented in table 9.5.

She has a Flexindex score of 54 points and exhibits hypermobility for her age, with scores over P_{99}. Mobility variability patterns are normal. Some joints have more marked hypermobility, particularly the ankle and elbow. Scores for all movements range from 2 to 4, with no sign of localized hypomobility.

These results show an overall hypermobility condition, which is often associated with mitral valve prolapse. In these cases, there is a tendency for movements primarily limited by ligaments to be more flexible. However, this cannot be seen in those older than 40 or 50 years of age. Her exceptional trunk flexion mobility is probably due to stretching training in previous years and to the smaller resistance of connective tissue to stretching, which is also a common feature in those with mitral valve prolapse. The score of 4 for movement XIV is quite unusual at this age, and perhaps this mobility should be reduced with specific muscle-strengthening exercises. It is suggested that current flexibility levels be maintained, without taking part in stretching classes.

Case Study 6

Case Study 6 is a 61-year-old man measuring 177 cm and weighing 96 kg who came to have an individual physical exercise program prescribed. His medical profile is suggestive of plurimetabolic syndrome. Even though he was very active in childhood and adolescence, he substantially decreased the practice of physical exercise over his adult life. He has no locomotor complaints and has started to walk for exercise over the past few months. The Flexitest data are presented in table 9.6.

The Flexindex value is somewhat lower than expected for age and gender, corresponding to a P_{35}. Variability indexes are normal, except for an increase in the distal–proximal index. In the joints, there is higher hypomobility in the hip, trunk, and shoulder as compared to other men.

Table 9.5									Case Study 5	
Movement	**I**	**II**	**III**	**IV**	**V**	**VI**	**VII**	**VIII**	**IX**	**X**
Movement score	4	3	3	2	3	2	3	2	4	2
Movement	**XI**	**XII**	**XIII**	**XIV**	**XV**	**XVI**	**XVII**	**XVIII**	**XIX**	**XX**
Movement score	3	2	2	4	3	3	2	2	3	2

Joint	Standardized score	
Ankle	70	
Knee	50	
Hip	50	
Trunk	60	
Wrist	40	
Elbow	70	
Shoulder	48	
Variability index	**Result**	
IMVI	0.71	
IJVI	0.53	
FEVI	1.27	
BSVI	1.08	
DPVI	1.18	
Flexindex	**Score**	**Percentile**
	54	99

There is significant hypomobility for trunk extension and relative hypomobility in all shoulder movements. On the other hand, mobility of the ankles and knees is preserved.

These results reflect a relatively common flexibility pattern in a middle-aged man who had been active, but now only does some walking and no specific stretching exercises other than a few repetitions of the classic movements of calf and thigh stretching before and after walking. His being overweight partially compromises the mobility of some movements, including hip adduction. Our recommendation is for him to focus more on stretching exercises, particularly extension movements of the trunk and shoulder.

Case Study 7

Case Study 7 is a 74-year-old man with a severe heart condition who measures 167 cm tall and weighs 78 kg. He was referred to our clinic for assessment and to attend a supervised exercise program. He has severe knee osteoarthritis, which partially impairs his gait. He is almost inactive physically, except for some daily exercises with free weights. The Flexitest data are presented in table 9.7.

Flexindex results reflect extreme hypomobility with a score of 9 out of 80 possible points, corresponding to a percentile of 2 for age and gender. Variability indexes show a heterogeneous mobility pattern with a high

Table 9.6									Case Study 6
Movement	**I**	**II**	**III**	**IV**	**V**	**VI**	**VII**	**VIII**	**IX**	**X**
Movement score	2	2	2	2	2	1	1	1	1	1
Movement	**XI**	**XII**	**XIII**	**XIV**	**XV**	**XVI**	**XVII**	**XVIII**	**XIX**	**XX**
Movement score	0	1	2	2	2	1	1	1	1	1

Joint	Standardized score	
Ankle	40	
Knee	40	
Hip	25	
Trunk	13	
Wrist	30	
Elbow	40	
Shoulder	20	
Variability index	**Result**	
IMVI	0.57	
IJVI	0.50	
FEVI	1.00	
BSVI	1.22	
DPVI	1.69	
Flexindex	**Score**	**Percentile**
	27	35

predominance of distal movements, variability reduction among the different movements, and also predominance of upper-limb mobility and in extension rather than flexion. The knee, hip, and shoulder joints exhibit extremely reduced mobility. Scores were either 0 or 1 for all movements.

These results reflect a pattern of severe hypomobility in an elderly person with an advanced heart condition who is practically physically inactive and also has problems in the locomotor system. Even though it is understandable that his osteoarthritis has caused major limitations in knee movements, the same is not true for the shoulders and hips, whose limited mobility is primarily due to lack of use. We recommend

incorporating a program of stretching training including active and passive executions up to the maximum range of motion, focusing on ankle, knee, and shoulder movements and trunk flexion, as an important part of his supervised exercise program.

Case Study 8

Case Study 8 is a 12-year-old girl with height of 161 cm and weighing 43 kg (body mass index of 16.6 kg\m^2). She has exercise-induced bronchospasm and was referred to us for assessment and advice on noncompetitive sports. She is still at a prepubertal stage (Tanner 2 to 3), and heart auscultation indicates mitral valve

Table 9.7								Case Study 7	
Movement	**I**	**II**	**III**	**IV**	**V**	**VI**	**VII**	**VIII**	**IX**	**X**
Movement score	0	1	0	0	1	0	0	0	0	1
Movement	**XI**	**XII**	**XIII**	**XIV**	**XV**	**XVI**	**XVII**	**XVIII**	**XIX**	**XX**
Movement score	1	1	1	1	1	1	0	0	0	0

Joint	Standardized score	
Ankle	10	
Knee	0	
Hip	5	
Trunk	13	
Wrist	20	
Elbow	20	
Shoulder	4	
Variability index	**Result**	
IMVI	0.50	
IJVI	0.36	
FEVI	1.33	
BSVI	0.45	
DPVI	2.81	
Flexindex	**Score**	**Percentile**
	9	2

prolapse. She has been dancing ballet and jazz regularly for some years and has no symptoms of the locomotor system. The Flexitest data are presented in table 9.8.

A 71-point Flexindex score reflects a clinical condition of hypermobility, regardless of age or gender. In addition, her percentile for the population of reference is over 99. Variability indexes show that her mobility pattern is excessively uniform for movements and for the different joints. Practically all joints have hypermobility values, except for the wrist. All individual scores were either 3 or 4, indicating that there is no restriction, even partial, for any movement.

These Flexindex and variability-index results match a clinical diagnosis of benign hypermobility. Performance in dancing is favored by hypermobility, particularly in classical ballet. If she is not really interested in a performance of excellence, a muscle-strengthening exercise program, carefully prescribed and gradually implemented, is recommended to prevent exaggerated movement arcs and early joint degeneration. Another alternative is to wait for the growth that takes place during puberty and the following years, when there is a natural tendency for hypermobility levels to be reduced.

Table 9.8   **Case Study 8**

Movement	I	II	III	IV	V	VI	VII	VIII	IX	X
Movement score	4	3	4	3	4	3	3	4	3	4

Movement	XI	XII	XIII	XIV	XV	XVI	XVII	XVIII	XIX	XX
Movement score	4	3	3	3	4	4	4	3	4	4

Joint	Standardized score	
Ankle	70	
Knee	70	
Hip	70	
Trunk	73	
Wrist	60	
Elbow	70	
Shoulder	76	

Variability index	Result	
IMVI	0.50	
IJVI	0.23	
FEVI	0.95	
BSVI	0.98	
DPVI	0.92	

Flexindex	Score	Percentile
	71	99

References

Abramson D, Roberts SM, Wilson PD. Relaxation of pelvic joints in pregnancy. *Surg Gynecol Obst* 1934;58:595-613.

Acasuso-Diaz M, Collantes-Estevez E. Joint hypermobility in patients with fibromyalgia syndrome. *Arthritis Care Res* 1998;11(1):39-42.

Acasuso-Diaz M, Collantes-Estevez E, Sanchez-Guijo P. Joint hyperlaxity and musculoligamentous lesions: Study of a population of homogeneous age, sex, and physical exertion. *Br J Rheumatol* 1993;32(2): 120-2.

Adrian M. Cinematographic, electromyographic, and electrogoniometric techniques for analyzing human movements. *Exerc Sports Sci Rev* 1973; 1:339-63.

Albee FH, Gilliland AR. Metrotherapy, or the measure of voluntary movement: Its value in surgical reconstruction. *JAMA* 1920;75:983-6.

Alexander RE, Battye CK, Goodwill CJ, Wash JB. The ankle and subtalar joints. *Clin Rheum Dis* 1982; 8:703-12.

Allander E, Bjornsson OJ, Olafsson O, et al. Normal range of joint movements in shoulder, hip, wrist and thumb with special reference to side: Comparison between two populations. *Int J Epidemiol* 1974;3: 253-61.

Alquier M. Un goniometre précis. *Rev Neurologique* 1916;23:515-6.

Alter MJ. *Science of flexibility.* Champaign, IL: Human Kinetics, 2nd edition, 1996.

Alter MJ. *Sport stretch.* Champaign, IL: Human Kinetics, 2nd edition, 1998.

American Academy of Pediatrics Committee on Sports Medicine. Atlantoaxial instability in Down syndrome. *Pediatrics* 1984;74:152-4.

American Alliance for Health, Physical Education, Recreation and Dance. *Physical best.* Reston, VA: AAHPERD, 1988.

American Association of Orthopaedics Surgeons. *Joint motion: Method of measuring and recording.* Edinburgh: Churchill Livingstone, 1965.

American College of Sports Medicine. The recom-mended quantity and quality of exercise for developing and maintaining cardio-respiratory and muscular fitness, and flexibility in healthy adults. *Med Sci Sports Exerc* 1998a; 30:975-91.

American College of Sports Medicine. Exercise and physical activity for older adults. *Med Sci Sports Exerc* 1998b;30(6):992-1008.

American College of Sports Medicine. *Guidelines for exercise testing and training.* Philadelphia: Lippincott Williams & Wilkins, 6th edition, 2000.

Anderson JAD, Sweetman BJ. A combined flexi-rule/hydrogoniometer for measurement of lumbar spine and its sagittal movement. *Rheumatol Rehabil* 1975;14:173-9.

Araújo CGS. Avaliação e treinamento da flexibilidade. In: Ghorayeb N, Barros Neto TL, editors. *O Exercício.* São Paulo: Atheneu, 1999a, p. 25-34.

Araújo CGS. Body flexibility profile and clustering among male and female elite athletes. *Med Sci Sports Exerc* 1999b;31(Suppl 5):S115 (abstract).

Araújo CGS. Correlação entre diferentes métodos lineares e adimensionais de avaliação da mobilidade articular. *Rev Bras Ciên Mov* 2000;8(2):25-32.

Araújo CGS. Existe correlação entre flexibilidade e somatotipo? Uma nova metodologia para um problema antigo. *Medicina do Esporte* 1983;7(3-4): 7-24.

Araújo CGS. Flexitest: An office method for evaluation of flexibility. *Sports & Medicine Today* 2001;1(2): 34-7.

Araújo CGS. Flexiteste: Proposição de cinco índices de variabilidade da mobilidade articular. *Rev Bras Med Esporte* 2002;8(1):13-9.

Araújo CGS. Flexiteste: Uma nova versão dos mapas de avaliação. *Kinesis* 1986;2(2):251-267.

Araújo CGS. *Medida e avaliação da flexibilidade: Da teoria à prática.* Universidade Federal do Rio de Janeiro, Instituto de Biofísica. Doctoral dissertation, 1987, 440 p.

Araújo CGS. Teste de sentar-levantar: Apresentação preliminar de um procedimento para avaliação em Medicina do Exercício e do Esporte. *Rev Bras Med Esporte* 1999c;5(5):179-82.

Araújo CGS, Haddad PCS. Efeitos do aquecimento ativo sobre a flexibilidade passiva. *Comunidade Esportiva* 1985;35:12-7.

Araújo CGS, Perez AJ. Características da flexibilidade em pré-escolares e escolares dos dois sexos. *Boletim da Federação Internacional de Educação Física* 1985;55(2):20-31.

Araújo CGS, Oliveira JA, Almeida MB. É válido utilizar versões condensadas do Flexiteste? *Rev Bras Ciên Mov* 2002;10(Suppl) (abstract) (in press).

Araújo CGS, Pereira MIR, Farinatti PTV. Body flexibility profile from childhood to seniority: Data from 1874 male and female subjects. *Med Sci Sports Exerc* 1998;30(Suppl 5):S115 (abstract).

Araújo DSMS, Araújo CGS. Auto-percepção das variáveis da aptidão física. *Rev Bras Med Esporte* 2002;8(2):37-49.

Arkkila PE, Kantola IM, Viikari JS. Limited joint mobility in non-insulin-dependent diabetic (NIDDM) patients: Correlation to control of diabetes, atherosclerotic vascular disease, and other diabetic complications. *J Diabetes Complications* 1997;11(4):208-17.

Armstrong AD, MacDermid JC, Chinchalkar S, Stevens RS, King GJ. Reliability of range-of-motion measurement in the elbow and forearm. *J Shoulder Elbow Surg* 1998;7(6):573-80.

Askling C, Lund H, Saartok T, Thorstensson A. Self-reported hamstring injuries in student-dancers. *Scand J Med Sci Sports* 2002;12(4):230-5.

Atha J, Wheatley DW. The mobilising effects of repeated measurement on hip flexion. *Br J Sports Med* 1976;10:22-5.

Bach DK, Green DS, Jensen GM, Savinar E. A comparison of muscular tightness in runners and nonrunners and the relation of muscular tightness to low back pain in runners. *J Orthop Sport Phys Ther* 1985;6:315-23.

Badley EM, Wood PHN. The why and wherefore of measuring joint movement. *Clin Rheum Dis* 1982;8:533-44.

Bandy WD, Irion JM, Briggler M. The effect of static stretch and dynamic range of motion training on the flexibility of the hamstrings muscles. *J Orthop Sports Phys Ther* 1998;27:295-300.

Bandy WD, Irion JM. The effect of time on static stretch on the flexibility of the hamstring muscles. *Phys Ther* 1994;74:845-50.

Barlow J, Benjamin B, Birt P, Hughes C. Shoulder strength and range-of-motion characteristics in bodybuilding. *J Strength Cond Res* 2002;16(3):367-72.

Barnett CH, Cobbold AF. Muscle tension and joint mobility. *Ann Rheum Dis* 1969;28:652-4.

Barrett CJ, Smerdely P. A comparison of community-based resistance exercise and flexibility exercise for seniors. *Aust J Physiother* 2002;48(3):215-9.

Baughman FA, Higgins JV, Wadswioeth TG, Demary MJ. The carrying angle in sex chromosome anomalies. *JAMA* 1974;230:718-20.

Beals RK. The normal carrying angle of the elbow: A radiographic study of 422 patients. *Clin Orthop* 1976;119:194-6.

Beighton P, Hóran F. Orthopaedic aspects of the Ehlers-Danlos syndrome. *J Bone Joint Surg* 1969;51B:444-53.

Beighton P, Hóran F. Dominant inheritance in familial generalised articular hypermobility. *J Bone Joint Surg* 1970;52B:145-59.

Beighton P, Solomon L, Soskolne CL. Articular mobility in an African population. *Ann Rheum Dis* 1973;32:413-8.

Bell BD, Hoshizaki TB. Relationships of age and sex with range of motion of seventeen joint actions in humans. *Can J Appl Sport Sci* 1981;6(4):202-6.

Benedetti A, Noacco C, Macor S, Pittaro I. Joint lesions in diabetes. *N Engl J Med* 1975;292:1033.

Benedetti A, Noacco C. Juvenile diabetic chiroarthropathy. *Acta Diabet Lat* 1976;13:54-67.

Bennell KL, Khan KM, Mathews BL, Singleton C. Changes in hip and ankle range of motion and hip muscle strength in 8-11 year old novice female ballet dancers and controls: a 12 month follow up study. *Br J Sports Med* 2001;35(1):54-9.

Bird HA, Brodie DA, Wright V. Quantification of joint laxity. *Rheumatol Rehabil* 1979;18:161-6.

Bird HA, Stowe J. The wrist. *Clin Rheum Dis* 1982;8:559-70.

Bird HA, Tribe CR, Bacon PA. Joint hypermobility leading to osteoarthrosis and chondrocalcinosis. *Ann Rheum Dis* 1978;37:203-11.

Biro F, Gewanter HL, Baum J. The hypermobility syndrome. *Pediatrics* 1983;72:701-6.

Birrell FN, Adebajo AO, Hazleman BL, Silman AJ. High prevalence of joint laxity in West Africans. *Br J Rheumatol* 1994;33(1):56-9.

Bohannon RW, Gajdosik R, LeVeau BF. Contribution of pelvic and lower limb motion to increases in the angle of passive straight leg raising. *Phys Ther* 1985;65:1501-4.

Bonci CM, Hensal FJ, Tiorg JS. A preliminary study on the measurement of static and dynamic flexibility at the glenohumeral joint. *Am J Sports Med* 1986;14(1):12-7.

Boone DC, Azen SP, Lin CM, Spence C, Barion C, Lee L. Reliability of goniometric measurements. *Phys Ther* 1978;58:1355-60.

Boone DC, Azen SP. Normal range of motion of joints in male subjects. *J Bone Joint Surg* 1979;61A:756-9.

Borms J, Van Roy P. Flexibility. In: Eston R, Reilly T, editors. Kinanthropometry and exercise physiology laboratory manual. London: E & FN Spon, 1996, p. 115-44.

Bosco JS, Gustafson WF. *Measurement and evaluation in physical education, fitness and sports.* Englewood Cliffs, NJ: Prentice Hall, 1983, p. 106-11.

Bouchard C, Shephard R, Stephens T, Sutton JR, McPherson BD. Exercise, fitness, and health: The Consensus statement. In: Bouchard C, Shephard R, Stephens T, Sutton JR, McPherson BD, editors. *Exercise, fitness, and health: A consensus of current knowledge.* Champaign, IL: Human Kinetics, 1990, p. 3-28.

Bower KD. The hydrogoniometer and assessment of glenohumeral joint motion. *Aust J Physiother* 1982;28:12-7.

Brach JS, VanSwearingen JM. Physical impairment and disability: relationship to performance of activities of daily living in community-dwelling older men. *Phys Ther* 2002;82:752-61.

Bressel E, McNair PJ. The effect of prolonged static and cyclic stretching on ankle joint stiffness, torque relaxation, and gait in people with stroke. *Phys Ther* 2002;82:880-7.

Bridges AJ, Smith E, Reid J. Joint hypermobility in adults referred to rheumatology clinics. *Ann Rheum Dis* 1992;51(6):793-6.

Brinkmann JR, Perry JV. Rate and range of knee motion during ambulation in healthy and arthritic subjects. *Phys Ther* 1985;65:1055-60.

Brodie DA, Bird HA, Wright V. Joint laxity in selected athletic populations. *Med Sci Sports Exerc* 1982;14:190-3.

Broer MR, Galles NR. Importance of relationship between various body measurements in performance of the toe-touch test. *Res Quart* 1958;29: 253-62.

Brown DA, Miller WC. Normative data for strength and flexibility of women throughout life. *Eur J Appl Physiol* 1998;78(1):77-82.

Buck CA, Damerion FB, Dow MJ, Skowlund HV. Study of normal range of motion in the neck utilizing a bubble goniometer. *Arch Phys Med Rehabil* 1959;40:390-2.

Buckwalter JA. Maintaining and restoring mobility in middle and old age: the importance of the soft tissues. *Instr Course Lect* 1997;46:459-69.

Bulbena A, Duro JC, Porta M, Martin-Santos R, Mateo A, Molina L, et al. Anxiety disorders in the joint hypermobility syndrome. *Psychiatry Res* 1993;46(1):59-68.

Burke DG, Culligan CJ, Holt LE. The theoretical basis of proprioceptive neuromuscular facilitation. *J Strength Cond Res* 2000;14(4):496-500.

Burley LR, Dobell HC, Farrell BJ. Relations of power, speed, flexibility, and certain anthropometric measures of high school girls. *Res Quart* 1961;32:443-8.

Calguneri M, Bird HA, Wright V. Changes in joint laxity occurring during pregnancy. *Ann Rheum Dis* 1982;41:126-8.

Campbell RR, Hawkins SJ, Maddison PJ, Reckless JPD. Limited joint mobility in diabetes mellitus. *Ann Rheum Dis* 1985;44:93-7.

Cantrell EF, Fisher T. The small joints of the hands. *Clin Rheum Dis* 1982;8:545-58.

Carter C, Sweetnam R. Familial joint laxity and recurrent dislocation of the patella. *J Bone Joint Surg* 1958;40B:664-7.

Carter C, Sweetnam R. Recurrent dislocation of the patella and of the shoulder: Their association with familial joint laxity. *J Bone Joint Surg* 1960;42B:721-7.

Carter C, Wilkinson J. Persistent joint laxity and congenital dislocation of the hip. *J Bone Joint Surg* 1964;46B:40-5.

Carter JEL. The somatotype of athletes: A review. *Hum Biol* 1970;42:535-69.

Carvalho ACG, Paula KC, Azevedo TMC, Nóbrega ACL. Relação entre flexibilidade e força muscular em adultos jovens de ambos os sexos. *Rev Bras Med Esporte* 1998;4(1):2-8.

Chandler TJ, Kibler WB, Uhl TL, Wooten B, Kiser A, Stone E. Flexibility comparisons of junior elite tennis players to other athletes. *Am J Sports Med* 1990;8(2):134-6.

Chang DE, Buschbacher LP, Edilich RF. Limited joint mobility in power lifters. *Am J Sports Med* 1988;16(3):280-4.

Chaves CPG, Araújo DSMS, Araújo CGS. Kinanthropometrical and clinical characteristics in adult women with mitral valve prolapse. *Med Sci Sports Exerc* 2001;33(Suppl 5):S75 (abstract).

Chaves CPG, Simão RF Jr, Araújo CGS. Ausência da variação da flexibilidade durante o ciclo menstrual em universitárias. *Rev Bras Med Esporte* 2002;8(6)212-8.

Chinn CJ, Priest JD, Kent BE. Upper extremity range of motion, grip strength, and girth in highly skilled tennis players. *Phys Ther* 1974;54:474-82.

Chung PK, Yuen CK. Criterion-related validity of sit-and-reach tests in university men in Hong Kong. *Percept Mt Skills* 1999;88:304-16.

Church JB, Wiggins MS, Moode FM, Crist R. Effect of warm-up and flexibility treatments on vertical jump performance. *J Strength Cond Res* 2001;15(3):332-6.

Clarke HH. Joint and body range of movement. *Physical Fitness Research Digest* 1975;5(4):1-21.

Clarkson HM. *Musculoskeletal assessment: Joint range of motion and manual strength.* Baltimore: Lippincott Williams & Wilkins, 1999.

Clayson SJ, Newman IM, Debevec DF, Anger RW, Skowlund HV, Kottke FJ. Evaluation of mobility of hip and lumbar vertebrae of normal young women. *Arch Phys Med Rehabil* 1962;43:1-8.

Clayson SJ, Mundale MO, Kottke FJ. Goniometer adaptation for measuring hip extension. *Arch Phys Med Rehabil* 1966;47:255-61.

Coelho CW, Araújo CGS. Relação entre aumento da flexibilidade e facilitações na execução de ações cotidianas em adultos participantes de programa de exercício supervisionado. *Revista Brasileira de Cineantropometria & Desempenho Humano* 2000;2(1):31-41.

Coelho CW, Velloso CR, Brasil RRLO, Vaisman M, Araújo CGS. Muscle power increases after resistance training in growth-hormone deficient adults. *Med Sci Sports Exerc* 2002;34(10)1577-81.

Cohen JL, Austin SM, Segal KR, Millman AE, Kim CS. Echocardiographic mitral valve prolapse in ballet dancers: a function of leanness. *Am Heart J* 1987;113(2 Pt 1):341-4.

Cole TM. Goniometry. In: Kottke FJ, Stillwell GK, Lehmann JF, editors. *Krusen's handbook of physical medicine and rehabilitation.* Philadelphia: WB Saunders, 3rd edition, 1982 p. 19-33.

Conwell HE. Flexo-extensometer. *Surg Gynecol Obst* 1925;40:710-1.

Coon V, Donato G, Houser C, Bleck EE. Normal ranges of hip motion in infants six weeks, three months and six months of age. *Clin Orthop* 1975;110:256-60.

Cooper Institute for Aerobic Research. *Fitnessgram: Test administration manual.* Champaign, IL: Human Kinetics, 2nd edition, 1999.

Corbin CB, Noble L. Flexibility: A major component of physical fitness. *Journal of Physical Education and Recreation* 1980;51(Jun):23-4, 57-60.

Corbin CB. Flexibility. *Clin Sports Med* 1984;3:101-17.

Cornbleet SL, Woolsey NB. Assessment of hamstring muscle length in school-aged children using the sit-and-reach test and the inclinometer measure of hip joint angle. *Phys Ther* 1996;76(8):850-5.

Cornwell A, Nelson AG, Sidaway B. Acute effects of stretching on the neuromechanical properties of the triceps surae muscle complex. *Eur J Appl Physiol* 2002;86(5):428-34.

Craib MW, Mitchell VA, Fields KB, Cooper TR, Hopewell R, Morgan DW. The association between flexibility and running economy in sub-elite male distance runners. *Med Sci Sports Exerc* 1996;28(6):737-43.

Cranney A, Goldstein R, Pham B, Newkirk MM, Karsh J. A measure of limited joint motion and deformity correlates with HLA-DRB1 and DQB1 alleles in patients with rheumatoid arthritis. *Ann Rheum Dis* 1999;58(11):703-8.

Cureton TK. Flexibility as an aspect of physical fitness. *Res Quart* 1941;12:381-90.

Danis CG, Mielenz TJ. Reliability of measuring active mandibular excursion using a new tool: The mandibular excursiometer. *J Orthop Sports Phys Ther* 1997;25(3):213-9.

Darcus HD, Salter N. The amplitude of pronation and supination with the elbow flexed to a right angle. *J Anat* 1953;87:169-84.

Decoster LC, Vailas JC, Lindsay RH, Williams GR. Prevalence and features of joint hypermobility among adolescent athletes. *Arch Pediatr Adolesc Med* 1997;151(10):989-92.

De Deyne PG. Application of passive stretch and its implications for muscle fibers. *Phys Ther* 2001;81:819-27.

De Felice C, Di Maggio G, Toti P, Parrini S, Salzano A, Lagrasta UE, Bagnoli F. Infantile hypertrophic pyloric stenosis and asymptomatic joint hypermobility. *J Pediatr* 2001;138(4):596-8.

De Inocencio J. Musculoskeletal pain in primary pediatric care: Analysis of 1000 consecutive general pediatric clinic visits. *Pediatrics* 1998;102(6):E63.

Denko CW, Boja B. Growth hormone, insulin, and insulin-like growth factor-1 in hypermobility syndrome. *J Rheumatol* 2001;28(7):1666-9.

Dequeker J. Benign familial hypermobility syndrome and Trendelenburg sign in a painting "The Three Graces" by Peter Paul Rubens (1577-1640). *Ann Rheum Dis* 2001;60(9):894-5.

De Vries HA. *Physiology of exercise: For physical education and athletics.* Dubuque, IA: Wm. C. Brown, 1974. p. 431-44.

Dickinson RV. The specificity of flexibility. *Res Quart* 1968;39:792-4.

Dockerty D, Bell RD. The relationship between flexibility and linearity measures in boys and girls

6-15 years of age. *J Hum Movem Studies* 1985;11:279-88.

Dorinson SM, Wagner ML. An exact technic for clinically measuring and recording joint motion. *Arch Phys Med Rehabil* 1948;39:468-75.

Dumas GA, Reid JG. Laxity of knee cruciate ligaments during pregnancy. *J Orthop Sports Phys Ther* 1997;26(1):2-6.

Dungy CI, Leupp M. Congenital hyperextension of the knees in twins. *Clin Pediatr* 1984;23:169-72.

Dunham WF. Ankylosing spondylitis: Measurement of hip and spine movements. *Br J Phys Med* 1949;12:126-89.

Duró JC, Vega E. Prevalence of articular hypermobility in schoolchildren: A one-district study in Barcelona. *Rheumatology* 2000;39:1153.

Duvall EN. Tests and measurements in physical medicine. *Arch Phys Med* 1948;389(4):202-5.

Einkauf DK, Gohdes ML, Jensen GM, Jewel MJ. Changes in spinal mobility with increasing age in women. *Phys Ther* 1987;67(3):370-5.

El-Garf AK, Mahmoud GA, Mahgoub EH. Hypermobility among Egyptian children: Prevalence and features. *J Rheumatol* 1998;25(5):1003-5.

Ellenbecker TS, Roetert EP, Piorkowski PA, Schulz DA. Glenohumeral joint internal and external rotation range of motion in elite junior tennis players. *J Orthop Sports Phys Ther* 1996;24(6):336-41.

Ellis MI, Burton KE, Wright V. A simple goniometer for measuring hip function. *Rheumatol Rehabil* 1979;18:85-90.

Ellis MI, Stowe J. The hip. *Clin Rheum Dis* 1982;8:655-76.

Ellison JB, Rose SJ, Sahrmann SA. Patterns of hip rotation range of motion: A comparison between healthy subjects and patients with low back pain. *Phys Ther* 1990;70(9):537-41.

Elward JF. Motion in the vertebral column. *Am J Roentgenol* 1939;42:91-9.

Escalante A, Lichtenstein MJ, Dhanda R, Cornell JE, Hazuda HP. Determinants of hip and knee flexion range: Results from the San Antonio Longitudinal Study of Aging. *Arthritis Care Res* 1999;12(1):8-18.

Escalante A, Lichtenstein MJ, Hazuda HP. Determinants of shoulder and elbow flexion range: Results from the San Antonio Longitudinal Study of Aging. *Arthritis Care Res* 1999;12(4):277-86.

Escalante A, Lichtenstein MJ, Hazuda HP. Walking velocity in aged persons: Its association with lower extremity joint range of motion. *Arthritis Rheum* 2001;45(3):287-94.

Etnyre BR, Abraham LD. Antagonist muscle activity during stretching: a paradox re-assessed. *Med Sci Sports Med* 1988;20(3):285-9.

Ewald HL, Rosenberg R, Mors NP. Panic anxiety, joint hypermobility and chromosome 15q changes. *Ugeskr Laeger* 2001 Nov 5;163(45):6291.

Fahey TD, Insel PM, Roth WT. *Fit & well: Core concepts and labs in physical education and wellness.* Mountain View, CA: Mayfield, 3rd edition, 1999.

Farinatti PTV, Soares PPS, Vanfraechem JHP. Influence de deux móis d'activités physiques sur la souplesse de femmes de 61 à 83 ans à partir d'un programme de promotion de la santé. *Sport* 1995;4:36-45.

Farinatti PTV, Araújo CGS, Vanfraechem JHP. Influence of passive flexibility on the ease for swimming learning in pre-pubescent and pubescent children. *Science et Motricité* 1997;31:16-20.

Farinatti PTV, Nóbrega ACL, Araújo CGS. Perfil da flexibilidade em crianças de 5 a 15 anos de idade. *Horizonte* [Lisboa]1998;14(82):23-31.

Fatouros IG, Taxildaris K, Tokmakidis SP, Kalapotharakos V, Agglelousis N, Athanasopoulos S, et al. The effects of strength training, cardiovascular training and their combination on flexibility of inactive older adults. *Int J Sports Med* 2002;23(2):112-9.

Feinstein AR. Multi-item "instruments" versus Virginia Apgar's principles of Clinimetrics. *Arch Intern Med* 1999;159:125-8.

Feland JB, Myrer JW, Schulthies SS, Fellingham GW, Measom GW. The effect of duration of stretching of the hamstring muscle group for increasing range of motion in people aged 65 years or older. *Phys Ther* 2001;81:1100-17.

Feldman DE, Shrier I, Rossingnol M, Abenhaim L. Risk factors for the development of low back pain in adolescence. *Am J Epidemiol* 2001;154:30-6.

Ferlic D. The range of motion of the normal cervical spine. *Johns Hopkins Hosp Bull* 1962;110:59-65.

Fieldman H. Effects of selected extensibility exercises on the flexibility of the hip joint. *Res Quart* 1966;37:326-31.

Finley FR, Karpovich PV. Electrogoniometric analysis of normal and pathological gaits. *Res Quart* 1964;35:379-84.

Finsterbush A, Pogrund H. The hypermobility syndrome: Musculosketal complaints in 100 consecutive cases of generalized joint hypermobility. *Clin Orthop* 1982;168:124-7.

Fisk GH. Some observations of motion at the shoulder joint. *Can Med Assoc J* 1944;50:213-6.

Fitzcharles MA, Duby S, Waddell RW, Banks E, Karsh J. Limitation of joint mobility (cheiroarthropathy) in adult non-insulin-dependent diabetic patients. *Ann Rheum Dis* 1984;43:251-7.

Fitzgerald GA, Greally JF, Drury MI. The syndrome of diabetes insipidus, diabetes mellitus and optic atrophy (DIDMOA) with diabetic cheiroarthropathy. *Postgrad Med J* 1978;54:815-7.

Fitzgerald GK, Wynvenn KJ, Rheault W, Rothschild B. Objective assessment with establishment of normal values for lumbar spinal range of motion. *Phys Ther* 1983;63:1776-81.

Fleckenstein SJ, Kirby RL, MacLeod DA. Effect of limited knee-flexion range on peak hip moments of force while transferring from sitting to standing. *J Biomech* 1988;21(11):915-8.

Fletcher GF, Balady G, Blair SN, Blumenthal J, Caspersen C, Chaitman B, et al. Statement on exercise: Benefits and recommendations for physical activity programs for all Americans: A statement for health professionals by the Committee on Exercise and Cardiac Rehabilitation of the Council on Clinical Cardiology, American Heart Association. *Circulation* 1996;94:857-62.

Forleo LH, Hilario MO, Peixoto AL, Naspitz C, Goldenberg J. Articular hypermobility in school children in São Paulo, Brazil. *J Rheumatol* 1993;20:916-7.

Fortier MD, Katzmarzyk PT, Malina RM, Bouchard C. Seven-year stability of physical activity and musculoskeletal fitness in the Canadian population. *Med Sci Sports Exerc* 2001;33:1905-11.

Fowles JR, Sale DG, MacDougall JD. Reduced strength after passive stretch of the human plantarflexors. *J Appl Physiol* 2000;89:1179-88.

Fowles JR, MacDougall JD, Tamopolsky MA, Sale DG, Roy BD, Yarasheski KE. The effects of acute passive stretch on muscle protein synthesis in humans. *Can J Appl Physiol* 2000;25(3):165-80.

Fredriksen H, Dagfinrud H, Jacobsen V, Maehlum S. Passive knee extension test to measure hamstring muscle tightness. *Scand J Med Sci Sports* 1997;7(5):279-82.

Frost M, Stuckey LA, Dorman G. Reliability of measuring trunk motions in centimeters. *Phys Ther* 1982;62:1431-7.

Gajdosik RL, Lusin G. Hamstring muscle tightness: Reliability of an active-knee-extension test. *Phys Ther* 1983;63:1085-8.

Gajdosik RL, LeVeau BF, Bohannon RW. Effects of ankle dorsiflexion on active and passive unilateral straight leg raising. *Phys Ther* 1985;65:1478-82.

Gedalia A, Person DA, Brewer EJ, Giannini EH. Hypermobility of the joints in juvenile episodic arthritis/arthralgia. *J Pediatr* 1985;107:873-6.

Gedalia A, Press J, Klein M, Buskila D. Joint hypermobility and fibromyalgia in schoolchildren. *Ann Rheum Dis* 1993;52(7):494-6.

Gersten JW, Ager C, Anderson K, Cenkovich F. Relation of muscle strength and range of motion to activities of daily living. *Arch Phys Med Rehabil* 1970;51:137-42.

Gilbert CB, Gross MT, Klug KB. Relationship between hip external rotation and turnout angle for the five classical ballet positions. *J Orthop Sports Phys Ther* 1998;27(5):339-47.

Gilliland AR. Norms for amplitude of voluntary movement (letter). *JAMA* 1921;77:1357.

Glanville AD, Kreezer G. The maximum amplitude and velocity of joint movements in normal male human adults. *Human Biol* 1937;9:197-211.

Gleim GW, Stachenfeld NS, Nicholas JA. The influence of flexibility on the economy of walking and jogging. *J Orthop Res* 1980;8(6):814-23.

Gleim GW, McHugh MP. Flexibility and its effects on sports injury and performance. *Sports Med* 1997;24(5):289-99.

Gogia PP, Braatz JH, Rose SJ, Norton BJ. Reliability and validity of goniometric measurements at the knee. *Phys Ther* 1987;67(2):192-5.

Goldberg B, Saraniti A, Witman P, Gavin M, Nicholas JA. Pre-participation sports assessment: An objective evaluation. *Pediatrics* 1980;736-45.

Goldman JA. Fibromyalgia and hypermobility. *J Rheumatol* 2001;28(4):920-1.

Grahame R. Joint hypermobility: Clinical aspects. *Proc Roy Soc Med* 1971;64:692-4.

Grahame R, Jenkins JM. Joint hypermobility: Asset or liability? A study of joint mobility in ballet dancers. *Ann Rheum Dis* 1972;31:109-11.

Grahame R, Edwards JC, Pitcher D, Gabell A, Harvey W. A clinical and echocardiographic study of patients with the hypermobility syndrome. *Ann Rheum Dis* 1981;40:541-6.

Grahame R, Pyeritz RE. The Marfan syndrome: joint and skin manifestations are prevalent and correlated. *Br J Rheumatol* 1995;34(2):126-31.

Grahame R. Joint hypermobility and genetic collagen disorders: Are they related? *Arch Dis Child* 1999;80:188-91.

Grahame R. Heritable disorders of connective tissue. *Baillieres Best Pract Res Clin Rheumatol* 2000a;14(2):345-61.

Grahame R. Hypermobility: Not a circus act. *Int J Clin Pract* 2000b;54(5):314-5.

Grahame R. Pain, distress and joint hyperlaxity. *Joint Bone Spine* 2000c;67(3):157-63.

Grahame R. Time to take hypermobility seriously (in adults and children). *Rheumatology* 2001;40(5):485-91.

Grahame R, Bird H. British consultant rheumatologists' perceptions about the hypermobility syndrome: A national survey. *Rheumatology* (Oxford) 2001;40(5):559-62.

Grana WA, Moretz JA. Ligamentous laxity in secondary school athletes. *JAMA* 1978;240:1975-6.

Gratacos M, Nadal M, Martins-Santos R, Pujana MA, Gago J, Peral B, Armengol L, Ponsa I, Miro R, Bulbena A, Estivill X. A polymorphic genomic duplication on human chromosome 15 is a susceptibility factor for panic and phobic disorders. *Cell* 2001;106(3):367-79.

Grgic A, Rosenbloom AL, Weber FT, Giordano B, Malone JI, Shuster JJ. Joint contracture: Common manifestation of childhood diabetes mellitus. *J Pediatr* 1976;88:584-8.

Grgic A, Rosenbloom AL, Weber FT, Giordano B. Joint contracture in childhood diabetes (letter). *N Engl J Med* 1975;292:372.

Gurewitsch AD, O'Neill MA. Flexibility of healthy children. *Arch Phys Ther* 1941;25(4):216-21.

Haas SS, Epps CH Jr, Adams JP. Normal ranges of hip motion in the newborn. *Clin Orthop* 1973;91:114-8.

Hahn T, Foldspang A, Vestergaard E, Ingermann-Hansen T. Active knee joint flexibility and sports activity. *Scand J Med Sci Sports* 1999;9(2):74-80.

Halbertsma JPK, van Bolhuis AI, Göeken LNH. Sport stretching: Effect on passive muscle stiffness of short hamstrings. *Arch Phys Med Rehabil* 1996;77:688-92.

Hall DM. Standardization of flexibility tests for 4-H Club members. *Res Quart* 1956;27:296-300.

Hall MG, Ferrell WR, Sturrock RD, Hamblen DL, Baxendale RH. The effect of the hypermobility syndrome on knee joint proprioception. *Br J Rheumatol* 1995;34(2):121-5.

Hamilton GF. Mobilization of the proximal interphalangeal joint: The influence of heat, cold, and exercise. *Phys Ther* 1967;47:1111-4.

Handler CE, Child A, Light ND, Dorrance DE. Mitral valve prolapse, aortic compliance, and skin collagen in joint hypermobility syndrome. *Br Heart J* 1985;54:501-8.

Harreby M, Nygaard B, Jessen T, Larsen E, Storr-Paulsen A, Lindahl A, et al. Risk factors for low back pain in a cohort of 1389 Danish school children: An epidemiologic study. *Eur Spine J* 1999;8(6):444-50.

Harris ML. A factor analytic study of flexibility. *Res Quart* 1969a;40:62-70.

Harris ML. Flexibility: Review of literature. *Phys Ther* 1969b;49:591-601.

Hartig DE, Henderson JM. Increasing hamstring flexibility decreases lower extremity overuse injuries in military basic trainees. *Am J Sports Med* 1999;27(2):173-6.

Hassoon A, Kulkarni J. Association between hypermobility and congenital limb deficiencies. *Clin Rehabil* 2002;16(1):12-5.

Hauer K, Rost B, Rütschle K, Opitz H, Specht N, Bärtsch P, Oster P, Schüerf G. Exercise training for rehabilitation and secondary prevention of falls in geriatric patients with a history of injurious falls. *J Am Geriatr Soc* 2001;49:10-20.

Heath BH, Carter JEL. A modified somatotype method. *Am J Phys Anthropol* 1967;27:57-74.

Hellebrandt FA, Duvall EN, Moore ML. The measurement of joint motion: Part III: Reliability of goniometry. *Phys Ther Rev* 1949;29:302-7.

Herbert RD, Gabriel M. Effects of stretching before and after exercising on muscle soreness and risk of injury: Systematic review. *Br Med J* 2002;325:468-72.

Hershler C, Milner M. Angle-angle diagrams in the assessment of locomotion. *Am J Phys Med* 1980a;59(3):109-25.

Hershler C, Milner M. Angle-angle diagrams in above-knee amputee and cerebral palsy gait. *Am J Phys Med* 1980b;59(4):165-83.

Hoffer MM. Joint motion limitation in newborns. *Clin Orthop* 1980;148:94-6.

Holland GJ. The physiology of flexibility: A review of the literature. *Kinesiol Rev* 1968;1:49-62.

Holt LE, Pelham TW, Burke DG. Modifications to the standard sit-and-reach flexibility protocol. *J Athletic Training* 1999;34(1):43-7.

Hóran FT, Beighton PH. Recessive inheritance of generalized joint hypermobility. *Rheumatol Rehabil* 1973;12:47-9.

Howe A, Thompson D, Wright V. Reference values for metacarpophalangeal joint stiffness in normals. *Ann Rheum Dis* 1985;44:469-76.

Hsieh CY, Walker JM, Gillis K. Straight-leg-raising test: Comparison of three instruments. *Phys Ther* 1983;63:1429-33.

Hubley CL, Kozey JW, Stanish WD. The effects of static stretching exercises and stationary cycling on range of motion at the hip joint. *J Orthop Sport Phys Ther* 1984;6:104-9.

Hudson N, Fitzcharles MA, Cohen M, Starr MR, Esdaile JM. The association of soft-tissue rheumatism and hypermobility. *Br J Rheumatol* 1998;37(4):382-6.

Hui SS, Yuen PY. Validity of the modified back-saver sit-and-reach test: A comparison with other protocols. *Med Sci Sports Exerc* 2000;32:1655-9.

Ingerval B. Variation of the range of movement of the mandible in relation to facial morphology in children. *Scand J Dent Res* 1970;78:535-43.

Jackson AW, Baker AA. The relationship of the sit and reach test to criterion measures of hamstring and back flexibility in young females. *Res Quart* 1986;57(5):183-6.

Jackson AW, Morrow JR Jr, Brill PA, Kohl HW III, Gordon NF, Blair SN. Relations of sit-up and sit-and reach tests to low back pain in adults. *J Orthop Sports Phys Ther* 1998;27(1):22-6.

Jessee EF, Owen DS, Sagar KB. The benign hypermobile joint syndrome. *Arthritis Rheum* 1980;23:1053-6.

Johns RJ, Wright V. Relative importance of various tissues in joint stiffness. *J Appl Physiol* 1962;17:824-8.

Johnson BL, Nelson JK. *Practical measurement and evaluation in physical education.* New York: Macmillan, 3rd edition, 1979. p. 78-93.

Johnson RP, Babbitt DP. Five stages of joint disintegration compared with range of motion in hemophilia. *Clin Orthop* 1985;201:36-42.

Jones AM. Running economy is negatively related to sit-and-reach test performance in international-standard distance runners. *Int J Sports Med* 2002;23(1):40-3.

Jones CJ, Rikli RE, Max J, Noffal G. The reliability and validity of a chair sit-and-reach test as a measure of hamstring flexibility in older adults. *Res Q Exerc Sport* 1998;69(4):338-43.

Jönhagen S, Nemeth G, Eriksson E. Hamstring injuries in sprinters: The role of concentric and eccentric hamstring muscle strength and flexibility. *Am J Sports Med* 1994;22(2):262-6.

Kadir N, Grayson MF, Goldberg AAJ, Swain MC. A new neck goniometer. *Rheumatol Rehabil* 1981;20:219-26.

Kapandji IA, Kandel MJ. *Physiology of the joints.* Edinburgh: Churchill Livingstone, 5th edition, 1997.

Kaplinsky C, Kenet G, Seligsohn U, Rechavi G. Association between hyperflexibility of the thumb and an unexplained bleeding tendency: Is it a rule of thumb? *Br J Haematol* 1998;101(2):260-3.

Karaaslan Y, Haznedaroglu S, Ozturk M. Joint hypermobility and primary fibromyalgia: A clinical enigma. *J Rheumatol* 2000;27(7):1774-6.

Karpovich PV, Karpovich GP. Electrogoniometer: A new device for study of joints in action. *Fed Proc* 1959;79.

Katzmarzyk PT, Craig CL. Musculoskeletal fitness and risk of mortality. *Med Sci Sports Exerc* 2002;34(5):740-4.

Katzmarzyk PT, Gledhill N, Pérusse L, Bouchard C. Familial aggregation of 7-year changes in musculoskeletal fitness. *J Gerontol Sci* 2001;56:B497-502.

Keats TE, Teeslink R, Diamond AE, Williams JH. Normal axial relationships of the major joints. *Radiology* 1966;87(5):904-7.

Kell RT, Bell G, Quinney A. Musculoskeletal fitness, health outcomes and quality of life. *Sports Med* 2001;31(12):863-73.

Kelliher MS. A report on the Kraus-Weber test in East Pakistan. *Res Quart* 1960;31(1):34-42.

Kendall HO, Kendall FP. Normal flexibility according to age groups. *J Bone Joint Surg* 1948;30A:690-4.

Kennedy L, Archer DB, Campbell SL, Beacom R, Carson DJ, Johnston PB. Limited joint mobility in type I diabetes mellitus. *Postgrad Med J* 1982;58:481-4.

Kettelkamp DB, Johnson RJ, Smidt GL, Chao EYS, Walker M. An electrogoniometric study of knee motion in normal gait. *J Bone Joint Surg* 1970;52A:775-90.

Key JA. Hypermobility of joints as a sex linked hereditary characteristic. *JAMA* 1927;88:1710-2.

Khan K, Roberts P, Nattrass C, Bennell K, Mayes S, Way S, et al. Hip and ankle range of motion in elite classical ballet dancers and controls. *Clin J Sport Med* 1997;7(3):174-9.

Khan KM, Bennell K, Ng S, Matthews B, Roberts P, Nattrass C, et al. Can 16-18-year-old elite ballet dancers improve their hip and ankle range of motion over a 12-month period? *Clin J Sports Med* 2000;10(2):98-103.

Kippers V, Parker AW. Toe-touch test: A measure of its validity. *Phys Ther* 1987;67(11):1680-4.

Kirby RL, Simms FC, Symington VJ, Garner JB. Flexibility and musculoskeletal symptomatology in female gymnasts and age-matched controls. *Am J Sports Med* 1984;18:143-8.

Kirk JA, Ansell BM, Bywaters EGL. The hypermobility syndrome: Musculoskeletal complaints associated with generalized joint hypermobility. *Ann Rheum Dis* 1967;26:419-25.

Klemp P. Hypermobility. *Ann Rheum Dis* 1997;56(10): 573-5.

Klemp P, Chalton D. Articular mobility in ballet dancers: A follow-up study after four years. *Am J Sports Med* 1989;17(1):72-5.

Klemp P, Learmonth ID. Hypermobility and injuries in a professional ballet. *Br J Sports Med* 1984;18: 143-8.

Klemp P, Williams SM, Stansfeld SA. Articular mobility in Maori and European New Zealanders. *Rheumatology* (Oxford) 2002;41(5):554-7.

Knudson DV, Magnusson P, McHugh M. Current issues in flexibility fitness. *The President's Council on Physical Fitness and Sports Research Digest* 2000;3(10): 1-8.

Kornberg M, Aulicino PL. Hand and wrist joint problems in patients with Ehlers-Danlos syndrome. *J Hand Surg* 1985;10A:193-6.

Kottke FJ, Mundale MO. Range of mobility of cervical spine. *Arch Phys Med Rehabil* 1959;40:379-82.

Kottke FJ. Therapeutic exercise to maintain mobility. In: Kottke FJ, Stillwell GK, Lehmann JF, editors. *Krusen's handbook of physical medicine and rehabilitation*. Philadelphia: WB Saunders, 3rd edition, 1982. p. 389-402.

Kraus H, Hirschland RP. Minimum muscular fitness test in school children. *Res Quart* 1954;25:178-88.

Krivickas LS, Feinberg JH. Lower extremity injuries in college athletes: relation between ligamentous laxity and lower extremity muscle tightness. *Arch Phys Med Rehabil* 1996;77(11):1139-43.

Kubo K, Kanehisa H, Kawakami Y, Fukunaga T. Influence of static stretching on viscoelastic properties of human tendon structures in vivo. *J Appl Physiol* 2001;90:520-7.

Kubo K, Kanehisa H, Fukunaga T. Effect of stretching training on the viscoelastic properties of human tendon structures in vivo. *J Appl Physiol* 2002;92: 595-601.

Kujala UM, Taimela S, Oksanen A, Salminen JJ. Lumbar mobility and low back pain during adolescence. A longitudinal three-year follow-up study in athletes and controls. *Am J Sports Med* 1997;25(3):363-8.

Kuo L, Chung W, Bates E, Stephen J. The hamstring index. *J Pediatr Orthop* 1997;17(1):78-88.

Kusinitz I, Keeney CE. Effects of progressive weight training on health and physical fitness of adolescent boys. *Res Quart* 1958;29:294-301.

Lankhorst GJ, Van De Stadt RJ, Van Der Korst JK. The natural history of idiopathic low back pain: A three-year follow-up study of spinal motion, pain and functional capacity. *Scand J Rehabil Med* 1985:17:1-4.

Larsson LG, Baum J, Mudholkar GS. Hypermobility: Features and differential incidence between the sexes. *Arthritis Rheum* 1987;30:1426-30.

Laubach LL, McConville JT. Relationships between flexibility, anthropometry, and the somatotype of college men. *Res Quart* 1966;37:241-51.

Lee EJ, Etnyre BR, Poindexter HB, Sokol DL, Toon TJ. Flexibility characteristics of elite female and male volleyball players. *J Sports Med Phys Fitness* 1989;29(1):49-51.

Lefevre J, Philippaerts RM, Delvaux K, Thomis M, Vanreusel B, Eynde BV, Claesseus AL, Lysens R, Renson R, Beunen G. Daily physical activity and physical fitness from adolescence to adulthood: A longitudinal study. *Am J Human Biol* 2000;12:487-97.

Lehnhard HR, Lehnhard RA, Butterfield SA, Beckwith DM, Marion SF. Health-related physical fitness levels of elementary school children ages 5-9. *Percept Mot Skills* 1992;75(3 Pt 1):819-26.

Leighton JR. A simple objective and reliable measure of flexibility. *Res Quart* 1942:13:205-16.

Leighton JR. Flexibility characteristics of males 10 to 18 years of age. *Arch Phys Med Rehabil* 1955;36: 571-8.

Leighton JR. An instrument and technic for the measurement of range of joint motion. *Arch Phys Med Rehabil* 1956;37(8):494-8.

Leighton JR. Flexibility characteristics of four specialized skill groups of college athletes. *Arch Phys Med Rehabil* 1957a:38:24-8.

Leighton JR. Flexibility characteristics of three specialized skill groups of champion athletes. *Arch Phys Med Rehabil* 1957b;38(9):580-3.

Len C, Ferraz MB, Goldenberg J, Oliveira LM, Araújo PP, Quaresma MR, et al. Pediatric Escola Paulista de Medicina Range of Motion Scale: A reduced joint count scale for general use in juvenile rheumatoid arthritis. *J Rheumatol* 1999;26(4): 909-13.

Lewkonia RM, Ansell BM. Articular hypermobility simulating chronic rheumatic disease. *Arch Dis Child* 1983;58:988-92.

Liemohn W, Martin SB, Pariser GL. The effect of ankle posture on sit-and-reach test performance. *J Strength Cond Res* 1997;11(4):239-41.

Lira VA, Araújo DSMS, Coelho CW, Araújo CGS. Sitting-rising test: Inter-observer reliability results. *Med Sci Sports Exerc* 1999;31(Suppl 5):S78 (abstract).

Livingstone B, Hirst P. Orthopedic disorders in school children with Down's syndrome with special reference to the incidence of joint laxity. *Clin Orthop Rel Res* 1986;207:74-6.

Looney MA, Plowman SA. Passing rates of American children and youth on the FITNESSGRAM criterion-referenced physical fitness standards. *Res Q Exerc Sport* 1990;61(3):215-23.

Loudon JK, Goist HL, Loudon KL. *Genu recurvatum* syndrome. *J Orthop Sports Phys Ther* 1998;27(5):361-7.

Lundbzaek K. Stiff hands in long-term diabetes. *Acta Med Scand* 1957;158:447-51.

Lysens R, Steverlynck A, van den Auweele Y, Lefevre J, Renson L, Claessens A, et al. The predictability of sport injuries. *Sports Med* 1984;1:6-10.

Macrae IF, Wright V. Measurement of back movement. *Ann Rheum Dis* 1969;28:584-9.

Madácsy L, Peja M, Korompay K, Biró B. Limited joint mobility in diabetic children: A risk factor of diabetic complications? *Acta Paediatr Hungarica* 1986;27(2):91-6.

Magnusson SP, Simonsen EB, Aagaard P, Sorensen H, Kjaer M. A mechanism for altered flexibility in human skeletal muscle. *J Physiol (Lond)* 1996;497(Pt 1):291-8.

Magnusson SP. Passive properties of human skeletal muscle during stretch maneuvers. *Scand J Med Sci Sports* 1998;8:65-77.

Magnusson SP, Julsgaard C, Aagaard P, Zacharie C, Ullman S, Kobayasi T, Kjaer M. Viscoelastic properties and flexibility of the human muscle-tendon unit in benign joint hypermobility syndrome. *J Rheumatol* 2001;28(12):2720-5.

Malcolm AD. Mitral valve prolapse associated with other disorders: Casual coincidence, common link, or fundamental genetic disturbance? (editorial) *Br Heart J* 1985;53:353-62.

Marks JS, Sharp J, Brear SG, Edwards JD. Normal joint mobility in mitral valve prolapse. *Ann Rheum Dis* 1983;42:54-5.

Maron BJ, Thompson PD, Puffer JC, McGrew CA, Strong WB, Douglas PS, et al. Cardiovascular preparticipation screening of competitive athletes: A statement for health professionals from the Sudden Death Committee and Congenital Cardiac Defects Committee – American Heart Association. *Circulation* 1996;94(4):850-6.

Martin JR, Ives EJ. Familial articular hypermobility and scapho-trapezial/trapezoid osteoarthritis in two siblings. *Rheumatology* 2002;41:1203-6.

Martin-Santos R, Bulbena A, Porta M, Gago J, Molina L, Duro JC. Association between joint hypermobility syndrome and panic disorder. *Am J Psychiatry* 1998;155(11):1578-83.

Massey BH, Chaudet NL. Effects of systematic, heavy resistive exercise on range of joint movement in young male adults. *Res Quart* 1956;27:41-51.

Mathews DK, Shaw V, Bohnen M. Hip flexibility of women as related to length of body segments. *Res Quart* 1957;28:352-6.

Mathews DK, Shaw V, Woods JB. Hip flexibility of elementary school boys as related to body segments. *Res Quart* 1959;30 297-302.

Mathews DK. *Measurement in physical education.* Philadelphia: WB Saunders, 5th edition, 1978. p. 357-65.

Maud PJ, Cortez-Cooper MY. Static techniques for the evaluation of joint range of motion. In: Maud PJ, Foster C, editors. *Physiological assessment of human performance.* Champaign, IL: Human Kinetics, 1995. p. 221-43.

McAtee R. *Facilitated stretching.* Champaign, IL: Human Kinetics, 2nd edition, 1999.

McHugh MP, Kremenic IJ, Fox MB, Gleim GW. The role of mechanical and neural restraints to joint range of motion during passive stretch. *Med Sci Sports Exerc* 1998;30:928-32.

McHugh MP, Connolly DA, Eston RG, Kremenic IJ, Nicholas SJ, Gleim GW. The role of passive muscle stiffness in symptoms of exercise-induced muscle damage. *Am J Sports Med* 1999;27(5):594-9.

McIntosh LJ, Stanitski DF, Mallett VT, Frahm JD, Richardson DA, Evans MI. Ehlers-Danlos syndrome: Relationship between joint hypermobility, urinary incontinence, and pelvic floor prolapse. *Gynecol Obstet Invest* 1996;41(2):135-9.

McMaster WC, Roberts A, Stoddard T. A correlation between shoulder laxity and interfering pain in competitive swimmers. *Am J Sports Med* 1998;26(1):83-6.

Merrit JL, McLean TJ, Erickson RP. Measurement of trunk flexibility in normal subjects: Reproducibility of three clinical methods. *Mayo Clin Proc* 1986;61:192-7.

Michels E. Measurement in physical therapy: On the rules for assessing numerals to observations. *Phys Ther* 1983:63:209-15.

Mikkelsson M, Salminen JJ, Kautiainen H. Joint hypermobility is not a contributing factor to musculoskeletal pain in pre-adolescents. *J Rheumatol* 1996;23(11):1963-7.

Mikkelsson M, Salminen JJ, Sourander A, Kautiainen H. Contributing factors of musculoskeletal pain in preadolescents: A prospective 1-year follow-up study. *Pain* 1998;77(1):67-72.

Moll JMH, Wright V. Normal range of spinal mobility: An objective clinical study. *Ann Rheum Dis* 1971;30:381-6.

Moll JMH, Wright V. The pattern of chest and spinal mobility in ankylosing spondilitis. *Rheumatol Rehabil* 1973;12:115-22.

Moller M, Ekstrand J, Oberg B, Gillquist J. Duration of stretching effect on range of motion in lower extremities. *Arch Phys Med Rehabil* 1985;66:171-3.

Moller MHL, Oberg BE, Gillquist J. Stretching exercise and soccer: Effect of stretching on range of motion in the lower extremity in connection with soccer training. *Int J Sports Med* 1985;6:50-2.

Moore MA, Hutton RS. Electromyographic investigation of muscle stretching techniques. *Med Sci Sports Med* 1980;12(5):322-9.

Moore ML. The measurement of joint motion: Part I: Introductory review of literature. *Phys Ther Rev* 1949a;29:195-205.

Moore ML. The measurement of joint motion: Part II: The technic of goniometry. *Phys Ther Rev* 1949b;29:256-64.

Morey MC, Pieper CF, Sullivan RJ Jr, Crowley GM, Cowper PA, Robbins MS. Five-year performance trends for older exercisers: A hierarchical model of endurance, strength, and flexibility. *J Am Geriatr Soc* 1996;44:1226-31.

Murray MP, Gore DR, Gardner GM, Mollinger LA. Shoulder motion and muscle strength of normal men and women in two age groups. *Clin Orthop* 1985;192:268-73.

Nattrass CL, Nitschke JE, Disler PB, Chou MJ, Ooi KT. Lumbar spine range of motion as a measure of physical and functional impairment: An investigation of validity. *Clin Rehabil* 1999;13(3):211-8.

Nef W, Gerber NJ. Hypermobility syndrome. When too much activity causes pain. *Schweiz Med Wochenschr* 1998;21;128(8):302-10.

Nelson AG, Guillory IK, Cornwell A, Kokkonen J. Inhibition of maximal voluntary isokinetic torque production following stretching is velocity-specific. *J Strength Cond Res* 2001;15(2):241-6.

Nelson AG, Kokkonen J. Acute ballistic muscle stretching inhibits strength performance. *Res Q Exerc Sport* 2001;72(4):415-9.

Nelson AG, Kokkonen J, Eldredge C, Cornwell A, Glickman-Weiss E. Chronic stretching and running economy. *Scand J Med Sci Sports* 2001;11:260-5.

Nelson JK, Johnson HL, Smith GC. Physical characteristics, hip flexibility and arm strength of female gymnasts classified by intensity of training across age. *J Sports Med Phys Fitness* 1983;23:95-101.

Nemethi CE. Normal wrist motions. *Indust Med Surg* 1953;22:230.

Nicholas JA. Injuries to knee ligaments: Relationship to looseness and tightness in football players. *JAMA* 1970;212:2236-9.

Nienaber CA, Von Kodolitsch Y. Therapeutic management of patients with Marfan syndrome: Focus on cardiovascular involvement. *Cardiol Rev* 1999;7(6):332-41.

Noer HR, Pratt DR. A goniometer designed for the hand. *J Bone Joint Surg* 1958;40A:1154-6.

Norkin CC, White DJ, White J. *Measurement of joint motion: a guide to goniometry.* Philadelphia: Davis, 1995.

Norton PA, Baker JE, Sharp HC, Warenski JC. Genitourinary prolapse and joint hypermobility in women. *Obstet Gynecol* 1995;85(2):225-8.

Nowak E. Angular measurements of foot-motion for application to the design of foot-pedals. *Ergonomics* 1972;15:407-15.

Noyes FR, Grood ES, Butler DL, Malek M. Clinical laxity tests and functional stability of the knee: Biomechanical concepts. *Clin Orthop* 1980;146:84-9.

Nygaard IE, Glowacki C, Saltzman CL. Relationship between foot flexibility and urinary incontinence in nulliparous varsity athletes. *Obstet Gynecol* 1996;87(6):1049-51.

Oberg B, Ekstrand J, Moller M, Gillquist J. Muscle strength and flexibility in different positions of soccer players. *Int J Sports Med* 1984;5:213-6.

O'Driscoll SL, Tomenson J. The cervical spine. *Clin Rheum Dis* 1982;8:617-30.

Ondrasik M, Rybar I, Rus V, Bosak V. Joint hypermobility in primary mitral valve prolapse patients. *Clin Rheumatol* 1988;7(1):69-73.

Pal B, Griffiths ID, Anderson J, Dick WC. Association of limited joint mobility with Dupuytren's contracture in diabetes mellitus. *J Rheumatol* 1987;14(3):582-5.

Parker AW, James B. Age changes in the flexibility of Down's syndrome children. *J Ment Defic Res* 1985;29(Pt 3):207-18.

Pate RR, Pratt M, Blair SN, Haskell WL, Macera CA, Bouchard C, et al. Physical activity and public health: A recommendation from the Centers for Disease Control and Prevention and the American College of Sports Medicine. *JAMA* 1995;273(5):402-7.

Pellecchia GL, Bohannon RW. Active lateral neck flexion range of motion measurements obtained with a modified goniometer: Reliability and estimates of normal. *J Manipulative Physiol Ther* 1998;21(7):443-7.

Penning L. Normal movements of the cervical spine. *Am J Roentgenol* 1978;130:317-26.

Pepin M, Schwarze U, Superti-Furga A, Byers PH. Clinical and genetic features of Ehlers-Danlos syndrome type IV, the vascular type. *N Engl J Med* 2000;342(10):673-80.

Phelps GS, Dickson JA. Spondylolisthesis and tight hamstrings. *J Bone Joint Surg* 1961;43A:505-12

Phillips DA, Hornak JE. *Measurement and evaluation in physical education.* New York: Wiley, 1979, p. 238-41.

Phillips M, Bookwalter C, Denman C, et al. Analysis of results from the Kraus-Weber test of minimum muscular fitness in children. *Res Quart* 1955;26: 315-23.

Pitcher D, Grahame R. Mitral valve prolapse and joint hypermobility: Evidence for a systemic connective tissue abnormality. *Ann Rheum Dis* 1982;41: 352-4.

Pollock ML, Franklin BA, Balady GJ, Chaitman BL, Fleg JL, Fletcher B, et al. Resistance exercise in individuals with and without cardiovascular disease: Benefits, rationale, safety, and prescription: An advisory from the Committee on Exercise, Cardiac Rehabilitation, and Prevention, Council on Clinical Cardiology, American Heart Association. *Circulation* 2000;101:828-33.

Potter P. The obliquity of the arm of the female in extension: The relation of the forearm with the upper arm in flexion. *J Anat Physiol* 1895;29:488-91.

Pountain G. Musculoskeletal pain in Omanis and the relationship to joint mobility and body mass index. *Br J Rheumatol* 1992;31(2):81-5.

Punzi L, Pozzuoli A, Pianon M, Bertazzolo N, Oliviero F, Scapinelli R. Pro-inflammatory interleukins in the synovial fluid of rheumatoid arthritis associated with joint hypermobility. *Rheumatology* 2001;40:202-4.

Qvindesland A, Jónsson H. Articular hypermobility in Icelandic 12-year-olds. *Rheumatology* 1999;38: 1014-6.

Rajapakse CN, Al-Orainey IO, Al-Harthi SS, Osman A, Smith J. Joint mobility and mitral valve prolapse in an Arab population. *Br J Rheumatol* 1987;26(6):442-4.

Rasmussen O, Tovborg-Jensen I. Joint range and deformity recorded by xerography. *Phys Ther* 1970;50:190-2.

al-Rawi Z, Nessan AH. Joint hypermobility in patients with chondromalacia patellae. *Br J Rheumatol* 1997;36(12):1324-7.

al-Rawi ZS, al-Rawi ZT. Joint hypermobility in women with genital prolapse. *Lancet* 1982;June 26: 1439-41.

Reid DC, Burnham RS, Saboe LA, Kushner SF. Lower extremity flexibility patterns in classical ballet dancers and their correlation to lateral hip and knee injuries. *Am J Sports Med* 1987;15(4): 347-52.

Reilly T. The concept, measurement and development of flexibility. In: Reilly T, editor. *Sports fitness and sport injuries.* London: Faber & Faber, 1981, p. 61-9.

Rejeski WJ, Brawley LR, Shumaker SA. Physical activity and health-related quality of life. *Exerc Sport Sci Rev* 1996;24:71-108.

Reynolds PMG. Measurement of spinal mobility: a comparison of three methods. *Rheumatol Rehabil* 1975;14:180-5.

Rezende AR, Faria AG Jr, Almeida LTP. Estudo descritivo sobre os índices de mobilidade articular da coluna vertebral, nos movimentos de inclinação lateral, num grupo de praticantes de natação. *Comunidade Esportiva* 1981;13:2-7.

Ricardo DR, Araújo CGS. Índice de massa corporal: Um questionamento científico baseado em evidências. *Arq Bras Cardiol* 2002;79(1)61-9.

Rikken-Bultman DG, Wellink L, van Dongen PW. Hypermobility in two Dutch school populations. *Eur J Obstet Gynecol Reprod Biol* 1997;73(2):189-92.

Rikli RE, Jones CJ. Assessing physical performance in independent older adults: Issues and guidelines. *J Aging Phys Activity* 1997;5:244-61.

Roaas A, Andersson GBJ. Normal range of motion of the hip, knee and ankle joints in male subjects, 30-40 years of age. *Acta Orthop Scand* 1982;53:205-8.

Rodriguez Y, Petersen F, Villareal A, Esquivel J, Reyes PA. Clinical features of idiopathic mitral valve prolapse. *Arch Inst Cardiol Mex* 1991;61(6):587-91.

Rosenbloom AL, Frias JL. Diabetes mellitus, short stature and joint stiffness: A new syndrome. *Clin Res* 1974;22:92 (abstract).

Rosenbloom AL, Malone JI, Yucha J, Van Cader TC. Limited joint mobility and diabetic retinopathy demonstrated by fluorescein angiography. *Eur J Pediatr* 1984;141:163-4.

Rosenbloom AL, Silverstein JH, Lezote DC, Richardson K, McCallum M. Limited joint mobility in childhood diabetes mellitus indicates increased risk for microvascular disease. *N Engl J Med* 1981;305: 191-4.

Rosenbloom AL, Silverstein JH, Lezote DC, Riley WJ, Maclaren NK. Limited joint mobility in diabetes mellitus of childhood: Natural history and relationship to growth impairment. *J Pediatr* 1982;101:874-8.

Rosenbloom AL, Silverstein JH, Riley WJ, MacLaren NK. Limited joint mobility in childhood diabetes: Family studies. *Diabetes Care* 1983;6:370-3.

Ross WD, Drinkwater DT, Bailey DA, Marshall GW, Leahy RM. Kinanthropometry: Definitions and traditions. In: Ostyn M, Beunen G, Simons J, editors. *Kinanthropometry II.* Baltimore: University Park Press, 1979, p. 3-32.

Rossi P, Fossaluzza V. Diabetic cheiroarthropathy in adult non-insulin-dependent diabetes. *Ann Rheum Dis* 1985;44:141-2.

Rozzi SL, Lephart SM, Gear WS, Fu FH. Knee joint laxity and neuromuscular characteristics of male and female soccer and basketball players. *Am J Sports Med* 1999;27(3):312-9.

Rusk HA. *Rehabilitation medicine.* St. Louis: Mosby; 1977, p. 9-15.

Russek LN. Hypermobility syndrome. *Phys Ther* 1999;79:591-9.

Sabari JS, Maltzev I, Lubarsky D, Liszkay E, Homel P. Goniometric assessment of shoulder range of motion: Comparison of testing in supine and sitting positions. *Arch Phys Med Rehabil* 1998;79(6):647-51.

Sady SP, Wortman M, Blanke D. Flexibility training: Ballistic, static or proprioceptive neuromuscular facilitation? *Arch Phys Med Rehabil* 1982;63:261-3.

Salter N. Methods of measurement of muscle and joint function. *J Bone Joint Surg* 1955;37B:474-91.

Sapega AA. Biophysical factors in range-of-motion exercise. *Phys Sportsmed* 1981;9(12):57-67.

Sapega AA, Nicholas J. The clinical use of musculoskeletal profiling in orthopedic sportsmedicine. *Phys Sportsmed* 1981;9(4):80-8.

Sarrafian SK, Melamed JL, Goshgarian GM. Study of wrist motion in flexion and extension. *Clin Orthop* 1977;126:153-9.

Saur PM, Ensink FB, Frese K, Seeger D, Hildebrant J. Lumbar range of motion: Reliability and validity of the inclinometer technique in the clinical measurement of trunk flexibility. *Spine* 1996;21(11): 1332-8.

Scharf Y, Nahir AM. Hypermobility syndrome mimicking juvenile chronic arthritis. *Rheumatol Rehabil* 1982;21:78-80.

Schenker AW. Goniometry: an improved method of joint motion measurement. *New York State J Med* 1956;56:539-45.

Schenkman M, Morey M, Kuchibhatla M. Spinal flexibility and balance control among community-dwelling adults with and without Parkinson's disease. *J Gerontol Sci* 2000;55:M441-5.

Schmidt GJ. Muscular endurance and flexibility components of the Singapore national physical fitness award. *Aust J Sci Med Sport* 1995;27(4):88-94.

Schnapf BM, Banks RA, Silverstein JH, Rosenbloom AL, Chesrow SE, Loughlin GM. Pulmonary function in insulin-dependent diabetes mellitus with limited joint mobility. *Am Rev Respir Dis* 1984;130:930-2.

Semine AA, Ertel NA, Goldberg MJ, Bull MJ. Cervical spine instability in children with Down syndrome. *J Bone Joint Surg* 1978;60A:649-52.

Seow CC, Chiow PK, Khong KS. A study of joint mobility in a normal population. *Ann Acad Med Singapore* 1999;28(2):231-6.

Shephard RJ, Berridge M, Montelpare W. On the generality of the "sit-and-reach" test: An analysis of flexibility data for an aging population. *Res Q Exerc Sport* 1990;61:326-30.

Shrier I. Stretching before exercise does not reduce the risk of local muscle injury: A critical review of the clinical and basic science literature. *Clin J Sport Med* 1999;9(4):221-7.

Shinabaerger NI. Limited joint mobility in adults with diabetes mellitus. *Phys Ther* 1987;67(2):215-8.

Siegler S, Lapointe S, Nobiline R, Berman AT. A six-degrees-of-freedom instrumented linkage for measuring the flexibility characteristics of the ankle joint complex. *J Biomech* 1996;29(7):943-7.

Silman AJ, Haskard D, Day S. Distribution of joint mobility in a normal population: Results of the use of fixed torque measuring devices. *Ann Rheum Dis* 1986;45:27-30.

Silva LPS, Palma A, Araújo CGS. Validade da percepção subjetiva na avaliação da flexibilidade de adultos. *Rev Bras Ciên Mov* 2000;8(3):15-20.

Silverman S, Costine L, Harvey W, Grahame R. Survey of joint mobility and in vivo skin elasticity in London schoolchildren. *Ann Rheum Dis* 1975;34: 177-80.

Simmons RW, Richardson C, Deutsch K. Limited joint mobility of the ankle in diabetic patients with cutaneous sensory deficit. *Diabetes Res Clin Pract* 1997;37(2):137-43.

Skinner JS, Baldini FD, Gardner AW. Assessment of fitness. In: Bouchard C, Shephard R, Stephens T, Sutton JR, McPherson BD, editors. *Exercise, health, and fitness.* Champaign, IL: Human Kinetics, 1990, p.109-19.

Smahel Z. Joint motion of the child hand. *Acta Chir Plast* 1975;17:113-24.

Smith AD, Stroud L, McQuen C. Flexibility and anterior knee pain in adolescent elite figure skaters. *J Pediatr Orthop* 1991;11(1):77-82.

Smith DS. Measurement of joint movement: An overview. *Clin Rheum Dis* 1982;8:523-32.

Smith JA. Relation of certain physical traits and abilities to motor learning in elementary school children. *Res Quart* 1956;27:220-8.

Smith LL, Brunetz MH, Chenier TC, McCammon MR, Houmard JA, Franklin ME, Israel RG. The effects of static and ballistic stretching on delayed onset muscle soreness and creatine kinase. *Res Q Exerc Sport* 1993;64(1):103-7.

Starkman H, Brink S. Limited joint mobility of the hand in type I diabetes mellitus. *Diabetes Care* 1982;5:534-6.

Starkman HS, Gleason RE, Rand LI, Miller DE, Soeldner JS. Limited joint mobility of the hand in patients with diabetes mellitus: Relation to chronic complications. *Ann Rheum Dis* 1986;45: 130-5.

Steel FLD, Tomlinson JDW. The "carrying angle" in man. *J Anat* 1958;92:315-7.

Stephens T, Sutton JR, McPherson BD, editors. *Exercise, health, and fitness.* Champaign, IL: Human Kinetics, 1990, p. 109-19.

Stoedefalke KG. Functional capacity testing. In: Ryan AJ, Allman FL Jr, editors. *Sports medicine.* New York: Academic Press, 1974, p. 447-62.

Storms H. A system of joint measurement. *Phys Ther Rev* 1955;35:369-71.

Stratford P, Agostino V, Brazeau C, Gowitzke BA. Reliability of joint angle measurement: A discussion of methodology issues. *Physiotherapy (Canada)* 1984;36:5-9.

Sturkie PD. Hypermobile joints in all descendants for two generations. *J Hered* 1941;32:232-4.

Sugahara M, Nakamura M, Sugahara K, et al. Epidemiological study on the change of mobility. *J Hum Ergol* 1981;10:49-60.

Suni JH, Mülunpalo SI, Asikainen TM, Laukkanen RT, Oja P, Pasanen ME, et al. Safety and feasibility of a health–related fitness test battery for adults. *Phys Ther* 1998;78(2):134-48.

Sutro CJ. Hypermobility of bones due to "over-lengthened" capsular and ligamentous tissues. *Surgery* 1947;21:67-76.

Taunton JE, Rhodes EC, Wolski LA, Donelly M, Warren J, Elliot J, McFarlane L, Leslie J, Mitchell J, Lauridsen B. Effect of land-based and water-based fitness programs on the cardiovascular fitness, strength and flexibility of women aged 65-75 years. *Gerontology* 1996;42:204-10.

Teitz CC. Sports medicine concerns in dance and gymnasts. *Clin Pediatr North Am* 1982;29:1517-42.

Thomas JS, Corcos DM, Hasan Z. The influence of gender on spine, hip, knee, and ankle motions during a reaching task. *J Motor Behav* 1998;30(2):98-103.

Tincello DG, Adams EJ, Richmond DH. Antenatal screening for postpartum urinary incontinence in nulliparous women: A pilot study. *Eur J Obstet Gynecol Reprod Biol* 2002;101(1):70-3.

Travers PR, Evans PG. Limitation of mobility in major joints of 231 sportsmen. *Br J Sports Med* 1976;10:35-6.

Tucci SM, Hicks JE, Gross EG, Campbell W, Danoff J. Cervical motion assessment: A new, simple and accurate method. *Arch Phys Med Rehabil* 1986;67:225-30.

Tully EA, Stillman BC. Computer-aided video analysis of vertebrofemoral motion during toe touching in healthy subjects. *Arch Phys Med Rehabil* 1997;78:759-66.

Tyler TF, Nicholas SJ, Campbell RJ, McHugh MP. The association of hip strength and flexibility with the incidence of adductor muscle strains in professional ice hockey players. *Am J Sports Med* 2001;29(2):124-8.

Tyler TF, Roy T, Nicholas SJ, Gleim GW. Reliability and validity of a new method of measuring posterior shoulder tightness. *J Orthop Sports Phys Ther* 1999;29(5):262-9.

Tyrance H. Relationships of extreme body types to ranges of flexibility. *Res Quart* 1958;29:349-59.

van Heuvelen MJG, Kempen GIJM, Ormel J, de Greef MHG. Self-reported physical fitness of older persons: A substitute for performance-based measures of physical fitness? *J Aging Phys Activity* 1997;5:298-310.

Veliskasis KP. Increased generalized ligamentous laxity in idiopathic scoliosis. *J Bone Joint Surg* 1973;55 A:435.

Verhoeven JJ, Tuinman M, Van Dongen PW. Joint hypermobility in African non-pregnant nulliparous women. *Eur J Gynecol Reprod Biol* 1999;82(1):69-72.

Vougiouka O, Moustaki M, Tsanaktsi M. Benign hypermobility syndrome in Greek schoolchildren. *Eur J Pediatr* 2000;159(8):628.

Wagner C. Determination of the rotary flexibility of the elbow joint. *Eur J Appl Physiol* 1977;37:47-59.

Wainerdi HR. An improved goniometer for arthrometry. *JAMA* 1952;149:661-2.

Walker BA, Beighton PH, Murdoch JKL. The marfanoid hypermobility syndrome. *Ann Intern Med* 1969;71:349-52.

Watson AWS. Factors predisposing to sports injury in school boy rugby players. *J Sports Med Phys Fitness* 1981;21:417-22.

Waugh KG, Minkel JL, Parker R, Coon VA. Measurement of selected hip, knee, and ankle joint motions in newborns. *Phys Ther* 1983;63:1616-21.

Weber S, Kraus H. Passive and active stretching of muscles: Spring stretch and control group. *Phys Ther Rev* 1949;29:407-10.

Weiss M. A multiple purpose goniometer. *Arch Phys Med Rehabil* 1964;45:197.

Wells KF, Dillon EK. The sit and reach: A test of back and leg flexibility. *Res Quart* 1952;23:115-8.

West CC. Measurement of joint motion. *Arch Phys Med* 1945;26(7):414-25.

Westling L, Mattiason A. General joint hypermobility and temporomandibular joint derangement in adolescents. *Ann Rheum Dis* 1992;51:87-90.

Wiechec FJ, Krusen FH. A new method of joint measurement and a review of the literature. *Am J Surg* 1939;43:659-68.

Wiesler ER, Hunter DM, Martin DF, Curl WW, Hoen H. Ankle flexibility and injury patterns in dancers. *Am J Sports Med* 1996;24(6):754-7.

Wiles P. Movements of the lumbar vertebrae during flexion and extension. *Proc Roy Soc Med* 1935;28:647-51.

Williams PO. Assessment of mobility in joints. *Rheumatism* 1957;13:13-6.

Wilmer HA, Elkins EC. An optical goniometer for observing range of motion of joints: A preliminary report of a new instrument. *Arch Phys Med* 1947;28(11):695-704.

Wilson GD, Statch WH. Photographic record of joint motion. *Arch Phys Med* 1945;26(6):361-2.

Wolf SL, Basmajian JV, Russe TC, Kutner M. Normative data on low back mobility and activity levels. *Am J Phys Med* 1979;58(5):217-29.

Wood ONH. Is hypermobility a discrete entity? *Proc Roy Soc Med* 1971;64:690-2.

Woods JY. Reliability of a non-invasive method for measuring back extension and defining normal ranges. *Phys Ther* 1985;65:675.

Wordsworth P, Ogilvie D, Smith R, Sykes B. Joint mobility with particular reference to racial variation and inherited connective tissue disorders. *Br J Rheumatol* 1987;26(1):9-12.

Wright V. Measurement of joint movement: Foreword. *Clin Rheum Dis* 1982;8:521-2.

Wright V, Hopkins R. The temporo-mandibular joint. *Clin Rheum Dis* 1982;8:715-22.

Wright V, Johns RJ. Observations on the measurement of joint stiffness. *Arth Rheum* 1960;3:328-40.

Wright V, Johns RJ. Quantitative and qualitative analysis of joint stiffness in normal subjects and in patients with connective tissue diseases. *Ann Rheum Dis* 1961;20:36-45.

Wynne-Davies R. Acetabular dysplasia and familial joint laxity: Two etiological factors in congenital dislocation of the hip: A review of 589 patients and their families. *J Bone Joint Surg* 1970;52B:704-16.

Wynne-Davies R. Hypermobility. *Proc Roy Soc Med* 1971;64:689-90.

Youdas JW, Garrett TR, Suman VJ, Bogard CL, Hallman HO, Carey JR. Normal range of motion of the cervical spine: An initial goniometric study. *Phys Ther* 1992;72(11):770-80.

Index

Note: The italicized *f* or *t* following a page number denote a figure or table on that page, respectively. The italicized *ff* or *tt* following a page number denotes multiple figures or tables on that page.

About the Author

Claudio Gil Soares de Araújo, MD, PhD, FACSM, is a professor at Gama Filho University and medical director of CLINIMEX in Rio de Janeiro, Brazil. A practicing physician in exercise and sports medicine who is dedicated to productive research, he has used Flexitest on more than 4,000 subjects in his practice since 1979.

Dr. Araújo has worked in supervised exercise programs and cardiopulmonary exercise testing since the 1980s. He also coordinated the medical evaluation of the Brazilian athletes during the 1988 and 1996 Olympic Games.

He earned his MD, MSc, and PhD from the University Federal Rio de Janeiro. As part of his medical training, he was a research fellow in cardiorespiratory and exercise areas in 1979 at McMaster University in Canada. Dr. Araújo is an American College of Sports Medicine (ACSM) fellow, committee member, and presenter at ACSM annual meetings.